Mountain Area Research and Management

Mountain Area Research and Management

Integrated Approaches

Edited by
Martin F. Price

London • Sterling, VA

First published by Earthscan in the UK and USA in 2007

ISBN: 978-1-84407-427-3

Typeset by FiSH Books, Enfield
Printed and bound in the UK by Antony Rowe, Chippenham
Cover design by Yvonne Booth

For a full list of publications please contact:

Earthscan
8–12 Camden High Street
London, NW1 0JH, UK
Tel: +44 (0)20 7387 8558
Fax: +44 (0)20 7387 8998
Email: earthinfo@earthscan.co.uk
Web: **www.earthscan.co.uk**

22883 Quicksilver Drive, Sterling, VA 20166-2012, USA

Earthscan publishes in association with the International Institute
for Environment and Development

A catalogue record for this book is available from the British Library

Library of Congress Cataloging-in-Publication Data

Mountain area research and management : integrated approaches / edited by
Martin F. Price.
    p. cm.
  ISBN-13: 978-1-84407-427-3 (hardback)
  ISBN-10: 1-84407-427-7 (hardback)
  1. Mountain ecology. 2. Ecosystem management. 3. Human ecology. 4.
Sustainable development. 5. Mountains—Research. I. Price, Martin F.
  QH541.5.M65M69 2007
  333.73—dc22

                                     2007009307

The paper used for this book is FSC-certified and totally chlorine-
free. FSC (the Forest Stewardship Council) is an international
network to promote responsible management of the world's
forests.

FSC
Mixed Sources
Product group from well-managed
forests and other controlled sources
Cert no. SGS-COC-2953
www.fsc.org
© 1996 Forest Stewardship Council

Contents

List of Boxes, Figures and Tables

Boxes

Figures

Tables

List of Contributors

Sandra Brown, Institute for Resources and Environment, University of British Columbia, Vancouver, British Columbia V6T 1Z3, Canada; sjbrown@interchange.ubc.ca

Tamsin Cooper, Institute of European Environmental Policy (IEEP), 28 Queen Anne's Gate, London SW1H 9AB, UK; TCooper@ieeplondon.org.uk

Danah Duke, Miistakis Institute, c/o Faculty of Environmental Design, University of Calgary, 2500 University Drive N.W., Calgary, Alberta T2N 1N4, Canada; dd@rockies.ca

Daniel B. Fagre, US Geological Survey (USGS) Northern Rocky Mountain Science Center, Glacier National Park, MT 59936, US; dan_fagre@usgs.gov

William Fisher, Parks Canada, Box 900, Banff, Alberta T1L 1K2, Canada

Roger Good, Environmental Consultant – Ecological Rehabilitation, 1178 Bungendore Road, Bungendore, NSW 2621, Australia; roger.good@bigpond.com.au

Guy Greenaway, Miistakis Institute, c/o Faculty of Environmental Design, University of Calgary, 2500 University Drive N.W., Calgary, Alberta T2N 1N4, Canada; guy@rockies.ca

Stuart Johnston, TransGrid, P.O. Box 139, Yass, NSW 2528, Australia; stuart.johnston@transgrid.com.au

Fidelis B. S. Kaihura, Agricultural Research and Development Institute, Ukiriguru, P.O. Box 1433, Mwanza, Tanzania; f.kaihura@yahoo.com

David J. Mattson, USGS Southwest Biological Science Center, Colorado Plateau Research Station, P.O. Box 5614, Northern Arizona University, Flagstaff, AZ 86011-5614, US; David.Mattson@nau.edu

Donald McKenzie, US Department of Agriculture Forest Service, Pacific Wildland Fire Sciences Lab, 400 N 34th Street #201, Seattle, WA 98103, US; donaldmckenzie@fs.fed.us

Troy Merrill, LTB Institute of Landscape Ecology, 208 South Main, Suite 7, Moscow, ID 83843, US; troy1@moscow.com

Bruno Messerli, Institute of Geography, University of Bern, Hallerstrasse 12, 3012 Bern, Switzerland; bmesserli@bluewin.ch

Paul Messerli, Institute of Geography, University of Bern, Hallerstrasse 12, 3012 Bern, Switzerland; mep@giub.unibe.ch

Paul Mitchell-Banks, Gartner Lee Limited, 6400 Roberts Street, Suite 490, Burnaby, BC, V5G 4C9, Canada; pmitchellbanks@gartnerlee.com

Jonathan Mitchley, Department of Agricultural Sciences, Imperial College London, Wye campus, High Street, Wye, Ashford, Kent TN25 5AH, UK; j.mitchley@imperial.ac.uk

Jeremias G. Mowo, Institut des Sciences Agronomiques du Rwanda (ISAR), B.P. 138, Butare, Rwanda; jgmowo@yahoo.com

David L. Peterson, Pacific Northwest Research Station, US Forest Service, 400 N 34th Street, Suite 201, Seattle, WA 98103, US; wild@u.washington.edu

Martin F. Price, Centre for Mountain Studies, Perth College-UHI, Crieff Road, Perth PH1 2NX, UK; martin.price@perth.uhi.ac.uk

Michael Quinn, Miistakis Institute, c/o Faculty of Environmental Design, University of Calgary, 2500 University Drive N.W., Calgary, Alberta T2N 1N4, Canada; quinn@ucalgary.ca

P. S. Ramakrishnan, School of Environmental Sciences, Jawaharlal Nehru University, New Delhi 110067, India; psr@mail.jnu.ac.in

Hans Schreier, Institute for Resources and Environment, University of British Columbia, Vancouver, British Columbia V6T 1Z3, Canada; star@interchange.ubc.ca

Riziki S. Shemdoe, University of Dar es Salaam, P.O. Box 31576, Dar es Salaam, Tanzania; rizikishemdoe@yahoo.com

Ann Stroud, African Highlands Initiative, P.O. Box 26416, Kampala. Sadly, Ann passed away while this book was in production. She had lived and worked in East Africa continuously since 1982 and will be missed by many friends and colleagues.

Joseph Tzanopoulos, Centre for Agri-Environmental Research (CAER), University of Reading, Earley Gate, P.O. Box 237, Reading, RG6 6AR, UK; j.tzanopoulos@reading.ac.uk

Clifford A. White, Parks Canada, Box 900, Banff, Alberta T1L 1K2, Canada; cliff.white@pc.gc.ca

Tomas Willebrand, Department of Forestry and Wildlife Management, Hedmark University College, Evenstad, 2480 Koppang, Norway; tomas.willebrand@hihm.no

List of Acronyms and Abbreviations

ACT	Australian Capital Territory
AHI	African Highlands Initiative
ALCES®	A Landscape Cumulative Effects Simulator
ANPC	Australian Network for Plant Conservation
ANU	Australian National University
ARM	adaptive resource management
ASARECA	Association for Strengthening Agricultural Research in East and Central Africa
ATV	all terrain vehicle
BAPPA/TIP	Beyond Agricultural Productivity to Poverty Alleviation/ Traditional Irrigation Project
BBVS	Banff Bow Valley Study
BC	British Columbia
BGC	BioGeoChemical (model)
BMP	best management practice
BNP	Banff National Park
C	carbon
CAD	conservation area design
CAP	Common Agricultural Policy
CBD	Convention on Biological Diversity
CCE	Crown of the Continent Ecosystem
CEA	cumulative effects assessment
CEM	Commission on Ecosystem Management
CGIAR	Consultative Group on International Agricultural Research
CIAT	Centro Internacional de Agricultura Tropical
CIMMYT	Centro Internacional de Mejaramiento de Maiz y Trigo
CIP	Centro Internacional de la Papa
CLIMET	Climate Landscape Interactions – Mountain Ecosystem Transect
CMP	Crown Managers' Partnership
CONDESAN	Consortium for the Sustainable Development of the Andean Ecoregion
CSIRO	Commonwealth Scientific and Industrial Research Organisation
DfID	Department for International Development
DIPNR	Department of Infrastructure, Planning and Natural Resources

DRD	Department of Research and Development
DRP	demographically robust population
ECA	East and Central Africa
EPA	Environment Protection Agency
ERP	evolutionarily robust population
EU	European Union
FRG	farmers' research group
GEF	Global Environment Facility
GIS	geographic information system
GMP	Global Mountain Programme
GTOS	Global Terrestrial Observing System
IAMC	Interagency Management Committee
IARC	international agricultural research centre
IBI	Integrated Benthic Index
ICIPE	International Centre for Insect Physiology and Ecology
ICRAF	International Centre for Research on Agroforestry
ICRISAT	International Crop Research Institute for the Semi-Arid Tropics
ICSU	International Council of Scientific Unions
IDR	interdisciplinary research
IDRC	International Development Research Centre
IFPRI	International Food Policy Research Institute
IGBP	International Geosphere-Biosphere Programme
IHDP	International Human Dimensions of Global Environmental Change Programme
IITA	International Institute for Tropical Agriculture
ILRI	International Livestock Research Institute
INRM	integrated natural resource management
IPGRI	International Plant Genetic Resources Institute
IRMMA	Interdisciplinary Research and Management in Mountain Areas
ISSC	International Social Science Council
IUCN	The World Conservation Union
IYM	International Year of Mountains
LAC	limits of acceptable change
LCV	League of Conservation Voters
LFA	less favoured area
LRMP	land and resource management plan
LUCC	Land-Use and Land-Cover Change
MAB	man and biosphere
M-KMA	Muskwa-Kechika Management Area
MRI	Mountain Research Initiative
N	nitrogen
NARI	national agricultural research institute

NCCR	National Centre of Competence in Research
NEPED	Nagaland Environmental Protection and Development
NGO	non-governmental organization
NPS	non-point source (of pollution)
NPWS	National Parks and Wildlife Service
NRM	natural resource management
NSW	New South Wales
NSWSF	NSW State Forests
NTFP	non-timber forest product
NUWA	National Urban Water Authority
PDO	Pacific Decadal Oscillation
PHABSIM	physical habitat simulation model
PLEC	People, Land Management and Environmental Change
PMC	Peace Managers Committee
PM&E	participatory monitoring and evaluation
POCD	policy-oriented conservation design
PRA	participatory rural appraisal
PTD	participatory technology development
PTDD	participatory technology development and dissemination
R&D	research and development
RHESSys	Regional Hydro-ecological Simulation System
RLAP	Regional Landscape Analysis Project
RTD	Research and Technological Development
SA	sustainability assessment
SAM	social accounting matrix
SDC	Swiss Agency for Development and Cooperation
SWE	snow water equivalent
TANESCO	Tanzania Electrical Supply Company
TEK	traditional ecological knowledge
UNCED	United Nations Conference on Environment and Development
UNESCO	United Nations Educational, Scientific and Cultural Organization
UNU	United Nations University
USAID	United States Agency for International Development
USDA	US Department of Agriculture
VDB	village development board
Y2Y	Yellowstone-to-Yukon

Integrated Approaches to Research and Management in Mountain Areas: An Introduction

Martin F. Price

This book is built around three major themes: research beyond disciplinary boundaries; integrated approaches to managing ecosystems; and mountain environments and the people who depend on them. This introductory chapter has two main aims. The first is to briefly present and discuss these three themes; the first two are clearly linked and provide the conceptual foundations for the book, while the third gives the setting for the four wide-ranging chapters and nine case studies that constitute the majority of the book. The second aim is to introduce these thematic chapters and case studies, which are further synthesized in the concluding chapter, which also presents the lessons learned from these experiences.

Research beyond disciplinary boundaries: Understanding a complex world

Recent decades have seen increases in collaboration between scientists from different disciplines within the scope of individual projects. In such projects, the degree of collaboration varies. At one end of a continuum of collaboration is multidisciplinary research, which according to Zube (1982):

> *usually consists of different disciplines investigating the same topic, but adhering to their traditional disciplinary languages and concepts. If integration is attempted it is frequently only in the form of an introduction, summary or conclusions to a report encompassing all the disciplinary conclusions.*

Further along the continuum is interdisciplinary research (COSEPUP, 2004):

> *a mode of research by teams or individuals that integrates informa-
> tion, data, techniques, tools, perspectives, concepts, and/or theories
> from two or more disciplines or bodies of specialized knowledge to
> advance fundamental understanding or to solve problems whose
> solutions are beyond the scope of a single discipline or field of
> research practice.*

Interdisciplinarity may also be divided into further components such as
pluridisciplinarity, crossdisciplinarity, and metadisciplinarity, though these
terms are used in various ways (Braun and Schubert, 2003; Thompson Klein,
1990). A number of authors have proposed rules for implementing interdisciplinary research; others argue that these may be helpful, but should not be
followed too rigorously, given that the development of any interdisciplinary
project is an intuitive process (Mackey, 2002; Szostak, 2002).

Bruce et al (2004), drawing on the work of Gibbons et al (1994), characterize two modes of interdisciplinary research. Mode 1 'brings together
researchers from different disciplines in order to overcome a blockage to
further development within a discipline, or to enable the discipline to move
into new and productive areas of research'. Mode 2 'addresses issues of social,
technical and/or policy relevance where the primary aim is problem-oriented
and discipline-related outputs are less central to the project design'. The
considerable increase in both modes means that the literature is growing
rapidly. For instance, within the natural sciences, from 1980 to 1999, the
number of publications whose title included the word multidisciplinarity or
interdisciplinarity doubled every seven years, a rate more than twice that of the
number of science journals (Braun and Schubert, 2003). The literature
comprises both work on interdisciplinary research *per se*, including books (for
example COSEPUP, 2004); Thompson Klein, 1990; Weingart and Stehr, 1999)
and journals (for example, *Issues in Integrative Studies*) as well as a plethora of
journals stressing the application of interdisciplinarity within a very wide range
of themes, often within what might be perceived as individual disciplines, such
as history, philosophy, neuroscience and mathematics. This underlines the fact
that, over time, interactions between individuals from different disciplines –
especially through 'Mode 1' interdisciplinary research – may lead to the evolution of new disciplines (see for example, Dogan and Pahre, 1990), so that
collaboration that would once have been regarded as interdisciplinary comes
to fall within the boundaries of one discipline. One example is landscape ecology (Wu, 2006).

Braun and Schubert (2003) identify two main reasons for the growth of
interdisciplinary research: first, many research problems are complex, thus
requiring the integration of perspectives from diverse disciplines; and second,
many underlying themes, such as chaos theory, are common to different disciplines. A further driver of interdisciplinary research is that of funding agencies,
which increasingly require that research projects should be interdisciplinary

and, in many cases, directed to developing knowledge that will be useful in addressing policy issues (see for example, Lowe and Phillipson, 2006; Tress et al, 2005). However, as Sperber (2003) notes, many grant proposals exhibit 'cosmetic interdisciplinarity': they 'have built in interdisciplinary rhetoric and describe future collaboration among people in different disciplines, but this is mostly done in order to meet the criteria for a grant'. At the other end of the process, reviewing the European Union's Fifth Framework Programme for Research and Technological Development (RTD), which explicitly focused on fostering interdisciplinary research, Bruce et al (2004) found:

> *disappointingly few projects ... that seemed ... to be clearly inter-disciplinary, particularly in terms of crossing the boundary between natural and social sciences... Even where projects were interdisci-plinary, the degree of interdisciplinarity varied. It tended to increase with time and with learning among partners.*

Most, though not all, authors agree that both multidisciplinary and interdisciplinary research primarily, if not exclusively, involves scientists, even though funding agencies increasingly require those who apply for grants within interdisciplinary programmes to specify how research teams will interact with other stakeholders, such as policy-makers and resource managers. This then links to an extension of the continuum of research collaboration into transdisciplinarity, a concept which emerged in the early 1970s (Balsiger, 2004). Definitions of transdisciplinarity differ (Lawrence and Després, 2004; Nicolescu, 2003; Thompson Klein, 2004). While some view transdisciplinarity as a further extension of interdisciplinarity, the key difference for the purposes of this book is that transdisciplinary research involves not only scientists, but other members of society, that is, it extends Mode 2 interdisciplinarity (Nowotny, 2003). A useful definition is 'a new form of learning and problem solving involving cooperation among different parts of society and academia in order to meet complex challenges of society' (Häberli et al, 2001). This definition was agreed at a major conference on transdisciplinarity in Switzerland in 2000, which was the final event of a major ten-year national research programme and subsequently led to the establishment of an internet-based network to advance transdisciplinary research (see www.transdisciplinarity.ch).

Although there is no consensus on a precise definition of transdisciplinarity, most authors do appear to agree that one reason that it is needed is because the challenges that our societies face are increasingly complex, resulting from interactions between at least four basic elements: economic agents, institutions, physical economic systems and ecosystems (Janssen, 2002). Transdisciplinary research is essential to address real-world problems, and is therefore context-specific (Hurni and Wiesmann, 2004; Thompson Klein, 2004). Thus, Pohl (2005) defines transdisciplinary research as research:

(a) *that takes into account the complexity of an issue – meaning the complex system of factors that together explain the issue's current state and its dynamic,*

(b) *that addresses both science's and society's diverse perceptions of an issue...*

(c) *that sets aside the idealised context of science in order to produce practically relevant knowledge...*

(d) *that deals with the issues and possible improvements of the status quo that are involved in balancing the diverse interests and inputs of individual stakeholders and disciplines.*

Importantly, transdisciplinary research does not replace disciplinary, multidisciplinary or interdisciplinary research; as shown in Figure 1.1, these are all essential elements of the transdisciplinary process, which also involves partners from outside academia.

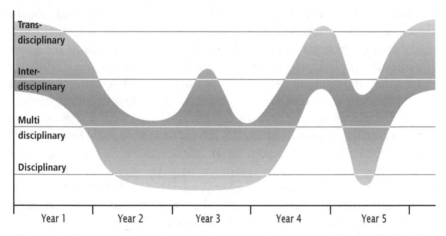

Figure 1.1 Emphasis of methods during the course of research involving specific disciplines and transdisciplinary approaches

Source: Hurni and Wiesmann (2004)

For any scientist, looking and working beyond the boundaries of his or her 'home' discipline is a challenge. Within academia, interdisciplinary and transdisciplinary research centres, institutions and programmes exist but they are relatively few (Pohl, 2005; Thompson Klein, 2004). A major report published by a committee of the US National Academies (COSEPUP, 2004) found that interdisciplinary collaboration 'has long been accepted and familiar in many industrial and government laboratories and other non-academic settings'. However, in academia, it is 'often impeded by administrative, funding and cultural barriers between departments', and hence the committee published a large number of recommendations directed at a wide range of audiences and aimed at facilitating interdisciplinary research. Similar comments and recommendations have also been made with regard to transdisciplinary research (for example, Thompson Klein et al, 2001; Thompson Klein, 2004). A key conclusion of much of this work is that, while institutions provide the context for either hindering or facilitating research beyond disciplinary boundaries, individual people also play critical roles, and committed leaders – who must also be good managers – are needed. Given that interdisciplinary and transdisciplinary research often involves PhD students, it should also be recognized that chances of completing dissertations may be poorer and the time for completion is often longer (Fry et al, 2005).

Between individuals, the key challenge to collaboration may not derive only from the world-view engendered by past training and experience in one or more disciplines or, in the case of transdisciplinarity, in the 'real world' outside academia. Moving beyond the natural/social science dichotomy – and many others – Pohl (2005) suggests that it may be more appropriate to group researchers according to how they relate to environmental problems. Thus, an 'Engaged Problem Solver' 'refuses to discuss a topic in an abstract way, and so is willing to think and debate issues only in a given context', while a 'Detached Specialist' is 'willing and able to discuss things in a much more abstract, generalised and context-free manner'. He concludes that researchers require several years of collaboration to gain respect for the other 'culture' before being able to develop and implement joint concepts; an issue explored by many of the papers in this book. The time required to develop respect, mutual understanding and effective methods of joint working implies a need for funding agencies to commit considerable resources, and for implementing institutions to provide the framework(s) to permit effective collaboration. To date, there have been relatively few examples, at least as far as academic institutions are concerned, though there have been some in the past, as discussed by Messerli and Messerli in this volume (and see also Thompson Klein et al, 2001); current examples include the Swiss National Centre of Competence in Research (NCCR), North-South (Hurni et al, 2004) and many of the activities supported by the Swedish Mistra Foundation (www.mistra-research.se; see Willebrand, this volume). The Networks of Excellence being developed through the European Commission's Sixth and Seventh Framework Programmes for RTD (Luukkonen et al, 2006) may also become effective mechanisms. All of these initiatives recognize that, because our world is increasingly complex, collaboration between disciplines and with diverse stakeholders is essential to address real-world current and future challenges. We hope that this book will support the contention that transdisciplinarity 'has become a major imperative across all sectors of society and knowledge domains, making it more than a fad or a fashion. It has become an essential mode of thought and action' (Thompson Klein, 2004).

Integrated approaches to managing ecosystems

To a significant extent, interdisciplinary and transdisciplinary research are driven by the interests of scientists who identify interesting issues and develop proposals for funding. At the same time, many of these issues are also the daily concerns of those who manage and use different natural resources, whether they are, for example, farmers, herders, foresters, gatherers of naturally growing plant species (for example, medicinal herbs, mushrooms, berries), or the managers of privately- or government-owned protected areas or forests. Particularly in mountain areas, a single individual may often have a number of these roles at one time or over the seasons of the year (Price and Thompson, 1997). The professionals who are concerned with the management of such resources – sometimes in collaboration with researchers based in academic institutions – have developed a wide range of integrated management methods. Though they may use, or be based on, interdisciplinary or transdisciplinary research, these methods implicitly go beyond this and recognize – albeit from a practical management rather than a research perspective – the need for diverse stakeholders to collaborate in order to address the realities of complex systems so as to ensure the sustainability of both livelihoods and ecosystems.

Some methods, particularly adaptive and/or ecosystem management, have primarily been developed in North America and formally implemented in the industrialized world (see for example, www.adaptivemanagement.net). Adaptive management (or adaptive environmental assessment and management) (Holling, 1978; Walters, 1986) originated in systems modelling, operations research and management, with stakeholders being involved both through workshops and as providers of information. Despite its roots in Canada, one of the earliest applications was in the Austrian Man and Biosphere (MAB) Programme, in the mountain village of Obergurgl when C. S. Holling, one of the originators of the approach, was working at the International Institute for Applied Systems Analysis (Himamowa, 1974). The most recent evolution of this approach has been through the Resilience Project (Berkes et al, 2003; Gunderson and Holling, 2002), which emphasized building resilience in periods of change in social-ecological systems and, as described by Folke et al (2003), has identified:

> *four factors that interact across temporal and spatial scales and that seem to be required for dealing with nature's dynamics in social-ecological systems:*
>
> - *learning to live with change and uncertainty;*
> - *nurturing diversity for reorganization and survival;*
> - *combining different types of knowledge for learning; and*
> - *creating opportunity for self-organization toward social-ecological sustainability.*

Ecosystem management is a broad concept developed within government agencies responsible for large land areas. It is based on many of the same principles as adaptive management and treated by some authors (for example, Berkes et al, 2003) as equivalent; others (for example, Norton, 2005) refer to adaptive ecosystem management. Ecosystem management encompasses a number of diverse themes, listed by Grumbine (1994, 1997) as hierarchical context, ecological boundaries, ecological integrity, data collection, monitoring, adaptive management, interagency cooperation, organizational change, humans embedded in nature and values. Any particular management agency and/or manager responsible for a particular location or landscape tends to aim at various goals associated with these themes, even though some goals may be in opposition to each other (Brunner and Clark, 1997). There are many definitions, probably not least because it has been adopted as an official policy by a number of federal agencies in the US (Grumbine, 1997; Haynes et al, 2005; Norton, 2005; Prato and Fagre, 2005), as well as Parks Canada (Zorn et al, 2001). Both adaptive management and ecosystem management are included in the six collaborative management methods for the management of complex human and natural systems compared by Blumenthal and Jannink (2000); the others are soft systems analysis (Checkland and Scholes, 1990), agroecosystem analysis (Conway and Barbier, 1990), rapid rural appraisal (McCracken et al, 1988), and participatory rural appraisal (Chambers, 1994). Blumenthal and Jannink (2000) usefully categorize these six methods, first, according to the spatial scale to which they have been applied, and their relative emphasis on simplifying natural resources and, second, according to the stages of management addressed and the attention given to social institutions.

While adaptive management has been applied in the developing world (for example, Berkes et al, 2003), agroecosystem analysis, rapid rural appraisal and participatory rural appraisal and another range of integrated methods have been both developed and implemented there, driven by the mandate of development agencies to both alleviate poverty and conserve the environment and, equally, by conservation organizations that also have an interest in providing benefits to poor people (Child, 2004; O'Riordan and Stoll-Kleemann, 2002; Pound et al, 2003; Stolton and Dudley, 1999; Zimmerer, 2006). Sayer and Campbell (2004) briefly summarize the sequence in which these methods appeared, starting with integrated rural development in the 1960s and continuing with integrated conservation and development projects in the 1970s. They identify eight more recent methods – integrated natural resource management, integrated catchment management, integrated water resource management, adaptive collaborative management, landscape management, multifunctional agriculture or forestry, and ecosystem approaches – and note that:

> *many claim that [these approaches] are attempts to put old wine into new bottles... The desire to achieve integration persists but our seeming inability to translate the theories of integration into*

practical achievements on the ground is leading to widespread disil-lusion. In frustration, we abandon one set of integrative buzzwords and replace them with others. What is surprising is not the improvement of integrative methods over the past 40 years – rather it is their fundamental similarity. The words have changed but the paradigm remains similar.

Nevertheless, Sayer and Campbell (2004) argue that considerable progress has been made in developing integrated tools and concepts, though they also recognize the challenges of developing and implementing them – which are largely similar to those of developing and implementing interdisciplinary or transdisciplinary research, as mentioned above. A key need, however, is to recognize that knowledge is always partial, and that adaptive approaches to management are essential. Sayer and Campbell (2004) write:

We never know enough about natural resource systems to manage them with certainty. Therefore, human interventions should always be experimental and should contribute to learning about the system... Integrated approaches to natural resources science will not yield precise recipes for managers, but they will help managers to make the right decisions and even more importantly to learn from their mistakes.

The principles of these various integrated approaches to management have been enshrined in the 12 principles of the ecosystem approach, 'a strategy for management of land, water and living resources that promotes conservation and sustainable use in an equitable way' (Convention on Biological Diversity, 2000) (see Box 1.1). It was adopted in 2000 by the Fifth Conference of the Parties to the Convention on Biological Diversity (Secretariat of the Convention on Biological Diversity, 2000), and is therefore binding on most of the world's nation states. The Commission on Ecosystem Management (CEM) of IUCN (The World Conservation Union) has taken a lead in refining these principles for practical implementation (Smith and Maltby, 2003), including their restructuring into the following five steps (Shepherd, 2004):

1 determining the main stakeholders, defining the ecosystem area, and developing the relationship between them;
2 characterizing the structure and function of the ecosystem, and setting in place mechanisms to manage and monitor it;
3 identifying the important economic issues that will affect the ecosystem and its inhabitants;
4 determining the likely impact of the ecosystem on adjacent ecosystems;
5 deciding on long-term goals and flexible ways of reaching them.

Box 1.1 Principles of the ecosystem approach

The principles of the ecosystem approach enshrined in the Convention on Biological Diversity (2000) are:

1   The objectives of management of land, water and living resources are a matter of societal choice.
2   Management should be decentralized to the lowest appropriate level.
3   Ecosystem managers should consider the effects (actual or potential) of their activities on adjacent and other ecosystems.
4   Recognizing potential gains from management, there is usually a need to understand and manage the ecosystem in an economic context. Any such ecosystem-management programme should:
    −   reduce those market distortions that adversely affect biological diversity;
    −   align incentives to promote biodiversity conservation and sustainable use;
    −   internalize costs and benefits in the given ecosystem to the extent feasible.
5   Conservation of ecosystem structure and functioning, to maintain ecosystem services, should be a priority target of the ecosystem approach.
6   Ecosystems must be managed within the limits of their functioning.
7   The ecosystem approach should be undertaken at the appropriate spatial and temporal scales.
8   Recognizing the varying temporal scales and lag-effects that characterize ecosystem processes, objectives for ecosystem management should be set for the long term.
9   Management must recognize that change is inevitable.
10 The ecosystem approach should seek the appropriate balance between, and integration of, conservation and use of biological diversity.
11 The ecosystem approach should consider all forms of relevant information, including scientific and indigenous and local knowledge, innovations and practices.
12 The ecosystem approach should involve all relevant sectors of society and scientific disciplines.

Source: Secretariat of the Convention on Biological Diversity (2000)

A recent CEM report (CEM, 2006), based on case studies on three continents, concludes that the ecosystem approach has great potential as a framework for planning, monitoring and analysis with regard to forces of change such as decentralization, urbanization, population growth, migration and climate change. However, it may need to be applied at different levels: 'national and sub national policy and legal frameworks may be just as important as what is going on within the ecosystem itself' and, most fundamentally, 'can only be fully applied where people are ready to share power and knowledge'.

In conclusion, the past four decades, and particularly the last, have been marked by increasing recognition among resource managers – from subsistence farmers to professional resource managers responsible for very large areas – that they are working in a complex world in which many systems

– ecological, social, economic, political, cultural – interact in increasingly uncertain ways and that decisions have to be made despite incomplete information. Many of the problems to be solved may be described as 'wicked' (Rittel and Webber, 1973), that is, they are not easily defined, there may be considerable disagreement between stakeholders as to what the problem is, and attempts to design a solution may change the problem or create a new one (see also Ludwig et al, 2001; Stewart et al, 2004; Quinn et al, this volume). The mechanistic, linear approaches to resource management that emerged in the 19th century and much of the 20th century have to give way to integrated, collaborative approaches; professional resource managers may have to be more humble, as 'midwives, not masters, in a timeless relationship dance with complex, diverse and dynamic human beings and earth ecosystems' (Kennedy and Koch, 2004).

Mountain environments and people

Mountains occupy 24 per cent of the global land surface (Kapos et al, 2000) and are home to 12 per cent of the global population (Huddleston et al, 2003). A further 14 per cent live adjacent to mountain areas (Meybeck et al, 2001); mountain people include not only poor and disadvantaged people in rural, often remote, communities, but also wealthy tourist communities and urban centres within and close to the mountains (including megacities such as Mexico City and Jakarta). Mountains, however, are important not only for people living in and adjacent to mountains, but for a very large proportion of the world's population of over 6.5 billion people.

The greatest value of mountain areas is that they are the source of up to 80 per cent of the world's fresh water. All the world's major rivers have their headwaters in the highlands, and more than half of humanity relies on this fresh water for domestic use, hydropower, industry, irrigation and transportation (Bandyopadhyay et al, 1997; Mountain Agenda, 1998; Viviroli et al, 2003). Mountain watersheds provide 19 per cent of the world's total electricity supply (Mountain Agenda, 2001; Schweizer and Preiser, 1997). Mountain forests play vital downstream protection roles by capturing and storing rainfall and moisture, maintaining water quality, regulating river flow, and reducing erosion and downstream sedimentation (Hamilton and Bruijnzeel, 1997). They also provide many millions of people with both timber and non-timber forest products (NTFPs) (for example, mushrooms, medicinal plants) (Price and Butt, 2000). The collection of NTFPs relates to one important aspect of the economic and health values of mountain biodiversity; mountains are 'hotspots' of biodiversity at both regional and global scales (Jeník, 1997; Körner and Spehn, 2002; Spehn et al, 2006). They harbour many endemic species, and are often sanctuaries for species eliminated from adjacent lowlands by human activity and/or changes in climate; equally, many mountain ranges function as biological corridors (Harmon and Worboys, 2004).

Mountain areas are also the original source of many food crops – including beans, corn, potatoes and wheat – often grown through the use of elaborate agricultural production systems and strategies based on altitudinal and ecological zonation (Grötzbach and Stadel, 1997). Individual mountain farmers may cultivate tens or even hundreds of varieties of different crops to ensure reliable sources of food in an uncertain environment; this agrodiversity may be of increasing value as the world's population continues to grow in an increasingly uncertain climate. Biological and cultural diversity are often closely inter-related, and mountains are also centres of cultural diversity (Grötzbach and Stadel, 1997; Stepp et al, 2005). Both of these linked types of diversity, as well as attractive landscapes, are among the key reasons why mountains are also major centres of tourism, the world's largest and fastest growing industry (Godde et al, 2000). Another value of many mountains for billions of people is their sacred or spiritual significance; pilgrims in their millions still visit sacred places (Bernbaum, 1997). However, while visitors can bring many positive economic benefits, there is also great potential for negative environmental and cultural consequences (Price et al, 1997).

Clearly, mountain areas are of global importance for many reasons. Yet, as late as the early 1990s, they were of interest to a relatively small number of scientists, development experts and decision-makers, as well as mountaineers. The United Nations Conference on Environment and Development (UNCED), held in Rio de Janeiro in 1992, presented a unique opportunity to move mountains onto the global stage, through the inclusion of a specific chapter in Agenda 21, the plan for action endorsed at UNCED by the heads of state or government of most of the world's nations (Stone, 2002). Chapter 13 of Agenda 21 is entitled 'Managing fragile ecosystems: sustainable mountain development', and includes two 'programme areas': generating and strengthening knowledge about the ecology and sustainable development of mountain ecosystems; and promoting integrated watershed development and alternative livelihood opportunities.

This chapter meant that, for the first time, mountains were accorded comparable priority in the global debate about environment and development with issues such as global climate change, desertification and deforestation. In 1998, the UN General Assembly re-emphasized the importance of the world's mountains by declaring that the year 2002 would be the International Year of Mountains (IYM). Thus, in the decade following UNCED, mountain issues were increasingly visible on the global agenda; and this status was further consolidated by diverse processes, activities and events during the IYM (Price and Hofer, 2005). Subsequently, mountain areas have remained on the global agenda through, for instance, the definition of a programme of work for mountain biodiversity under the Convention on Biological Diversity (Anon, 2004) and the inclusion of a chapter on mountain systems in the Millennium Ecosystem Assessment (Körner and Ohsawa et al, 2005).

Over the 15 years since UNCED, the diverse global values of mountain areas have been frequently stated in both academic and policy documents, often in association with statements recognizing that a high proportion of mountain people are vulnerable to food shortages and have inequitable access to resources and political power; that a disproportionate number of wars and conflicts take place in mountain areas; and that such challenges to the well-being of mountain people are further accentuated by the dynamic nature of mountain environments (and hence the high frequency of natural hazards) and climate change (Hurni, 2003; Körner and Ohsawa et al, 2005; Messerli and Ives, 1997; Rhoades, 1997; Royal Swedish Academy of Sciences, 2002; Stone, 1992; United Nations, 2002). A further theme of such documents is uncertainty, both for those making decisions about how to make a living in and from mountain environments and for those who seek to understand them from scientific perspectives – and it is increasingly recognized that traditional or indigenous knowledge may often be at least as important as knowledge based on 'western' science (Ives et al, 1997; Ramakrishnan, this volume). Mountain environments, and the societies that live in and depend on them, are clearly complex, and hence there are strong needs for integrated approaches to both research and management.

Identifying the case studies

The immediate genesis of this book was a conference on 'Interdisciplinary Research and Management in Mountain Areas' (IRMMA), which took place at the Banff Centre, Banff, in Canada from 23 to 27 September 2004. This was the fourth in a series of five 'Mountain Communities' conferences organized by Mountain Culture at the Banff Centre (see www.banffcentre.ca/mountainculture). The proceedings of this conference (Taylor and Ryall, 2004) provide an important complement to this book, as they include records of the discussions of the opening and concluding plenary sessions as well as abstracts and short papers that are not included here.

Drawing on many of the principles of integrated research and management discussed above, the process of choosing case studies for presentation – and then inclusion in this book – was unusually rigorous. The steering committee (Micheline Manseau, Martin Price, Jorge Recharte, Jillian Roulet, Leslie Taylor, Cliff White) and programme committee (James Gardner, Marty Magne, Ana Maria Ponce, Martin Price, Cliff White) identified about 50 projects from mountain areas around the world that were described in some way as interdisciplinary. The managers or proponents of these projects were sent a questionnaire designed principally to evaluate the degree to which interdisciplinarity or transdisciplinarity had been considered in developing the project, as follows:

- aims/objectives: how these were defined, and what groups were involved;
- interdisciplinarity: disciplines represented in the project team; mechanisms used to foster interdisciplinary concepts and working; principal challenges to interdisciplinary working;

- stakeholder involvement: groups of stakeholders involved in the project and how (definition of issues, priorities and activities; implementation of research; dissemination of results/outcomes);
- project outcomes: main results/outcomes of the research; means of disseminating these; principal publications; influence of the project on local management decisions, local/regional policies, national policies and other projects.

The responses to the questionnaire were then scored, using a five-point scale, by the programme committee, to assess the extent to which:

- project aims and objectives committed to an interdisciplinary approach;
- the project team was interdisciplinary and well qualified;
- non-scientific stakeholders (for example, local communities, decision-makers, planners) were involved in developing and implementing the project;
- innovative or leading-edge mechanisms were used to foster interdisciplinarity;
- challenges/lessons learned were clearly defined;
- the project had succeeded in meeting its objectives (if completed) or appeared likely to succeed (if still under way);
- the outcomes indicated that an interdisciplinary approach had been successful;
- the project had clear effects on management and policy in the target area;
- the project had valuable outcomes outside the project area;
- the project appeared to contain elements of planning and implementation that could be readily transferable to other mountain areas.

As the intention was to hold a truly global conference, the proponents of the highest scoring projects from six regions – Africa, Asia, Europe, Latin America, Northern America, Oceania – were invited to the conference. Almost all of these case studies are included in this book. In order to facilitate comparison between chapters, each author was asked to prepare a chapter according to an outline that generally followed the themes in the original questionnaire. All of these chapters have been revised, most recently in mid- to late 2006.

The chapters

Following this introduction are four wide-ranging chapters: two describing decades-long experiences of interdisciplinary and transdisciplinary research; and two with thematic foci. These are followed by nine case studies from four continents, and a concluding chapter. The authors, who include employees of government agencies and research institutes, consultants and university academics, have their roots in a wide range of disciplines.

The first chapter, by Bruno Messerli and Paul Messerli, takes a very broad view of interdisciplinary and transdisciplinary research in mountain areas over the past two centuries, starting with the work of Alexander von Humboldt. A major force for the development of integrated approaches was UNESCO's (United Nations Educational, Scientific and Cultural Organization's) Man and the Biosphere (MAB) Programme, within which, starting in 1979, the two authors successively led the Swiss component. However, though the model created in the Swiss Alps effectively contributed to the evolution of scientific practice and understanding and contributed to policy development in Switzerland, it needed significant adaptation to be useful in the mountains of the developing world, where the driving forces are often very different. This led to transdisciplinary approaches and to 'syndrome mitigation research' as mentioned above (Hurni and Wiesmann, 2004). At the same time, elements of the Swiss MAB model remain in current research programmes on global change, which must be both interdisciplinary and transdisciplinary despite the challenges of collaboration and of differing priorities in industrialized and developing countries.

Messerli and Messerli conclude that global change and globalization may lead to the disappearance of culture-specific land-use systems, which are the focus of the second chapter, by P. S. Ramakrishnan. His paper focuses on the '*jhum*' (slash-and-burn) shifting agriculture system practised in northeast India. Government agencies have viewed this system as 'primitive' for a century or more. Yet, more than three decades of interdisciplinary research and, increasingly, transdisciplinary research involving local farmers who are the holders of traditional ecological knowledge (TEK) have shown that this highly complex multi-cropping system is well adapted to a wide variety of situations; adapted versions can even sustain livelihoods when the rotational cycle crops below five years. *Jhum* is also linked to other land-use systems that ensure the self-sufficiency of village ecosystems within their cultural landscape. The outcomes of the decades of scientific research – and generations of TEK – described in the paper are now being used to inform development initiatives in much of northeast India.

Following these two chapters that focus primarily on integrated approaches to research are two that focus specifically on integrated approaches to management. Both address issues that go well beyond mountain environments, though they also give particular attention to these environments. David Mattson and Troy Merrill note that the development of policies for the conservation of species and ecosystems should be based on finding common ground, taking into consideration differing myths and value orientations, as well as differences between national and regional goals and preferences. They also note the many complexities and uncertainties that lead to a need for adaptive management, rather than prescriptive 'tools' such as conservation area design – but that most conservation agencies and organizations learn poorly and are unresponsive to information that questions the status quo. Nevertheless, conservation is fundamentally about changing human behaviour; that is, it is about decision-making and policy. Mattson and Merrill therefore propose that conservation design should be problem-oriented, based on the concept of 'sufficient intelligence'. Their resulting approach is termed policy-oriented conservation design, and they apply this to the specific issue of grizzly bears in the Rocky Mountains of the western US.

In the last wide-ranging chapter, Sandra Brown and Hans Schreier address the need for innovative watershed management, drawing on experience from around the world. Importantly, they recognize that, while most of the area of mountain watersheds is rural, many mountain people live in urban environments and depend on the water that flows through these. In a world with a growing population and a changing and increasingly variable climate, more careful management of water sources and of demand for water is essential to ensure that as much of the water falling in mountain areas can be efficiently converted to food and fibre. In some cases, this may include the reconversion of the dams and channelized systems that have been developed in many mountain regions. They conclude that adaptive management approaches are essential, involving integrated approaches in which agencies work together across the traditional sectors – and also with the public – to ensure over the coming decades the long-term availability of appropriate quantities of high-quality water, the key resource from mountain areas.

The first two case studies are from densely populated mountain areas in Tanzania and represent work carried out within international programmes. Jeremias Mowo, Riziki Shemdoe and Ann Stroud present work in the Usumbara mountains within the African Highlands Initiative, the East African component of the Global Mountain Programme of the Consultative Group on International Agricultural Research (CGIAR). The project area is characterized by many problems, including resource degradation, poor market infrastructure, poorly formulated policies and land scarcity. While there were several challenges to interdisciplinary work, participatory monitoring and evaluation showed that the project scientists increasingly worked together as a team and with local farmers and were able to develop various methodologies for improving systems and landscapes, research and development strategies, and institutional capacity for sustainable development. The project from Mount Meru, presented by Fidelis Kaihura, is part of the global People, Land Management and Environmental Change (PLEC) project, supported by the United Nations University (UNU) and the Global Environment Facility (GEF). As in other PLEC projects, the project area was chosen because of its critical ecosystems and important biodiversity; work focused on small-scale, intensively managed agricultural systems, with a particular focus on starting from existing good agricultural practice. Many different groups of stakeholders were involved throughout the project, from defining project priorities and activities to implementing research and disseminating results. As in the Usumbara mountains, the development of interdisciplinary practices took a while to develop; as did recognition that farmers have much valuable knowledge and undertake their own experiments. For both projects, results and outcomes have been disseminated at many levels, through means ranging from farmer-to-farmer discussion and demonstration to local, regional and national meetings and scientific publications.

The next two case studies are from Europe. In Sweden, the Mistra Foundation has funded a number of multi-year interdisciplinary programmes focusing on strategic long-term environmental issues. One of these programmes, described by Tomas Willebrand, has been in the mountains of northern Sweden. This region has a low (and declining) population, yet there are various conflicts over the ownership, use and management of resources such as forests (used both for reindeer grazing and forestry), fish and game, as well as over the appropriate number of large carnivores and the development of tourism and recreation. These conflicts involve both local people and others from outside the region. The goal of the Mountain Mistra programme has been to develop strategies for the management and long-term development of the region's resources based on increased scientific understanding bringing together the viewpoints of diverse natural and social scientific disciplines. Following a first phase in which research by scientists from different disciplines was not very well integrated, the second phase began with extensive discussions between researchers and stakeholders. This led to a complete restructuring of the programme, with new emphases on interdisciplinary and transdisciplinary work and an adaptive approach to scenario development. This was underpinned by a strong analytical framework and complemented by a highly developed communications strategy, coordinated by a full-time communicator.

The BioScene project, the subject of the chapter by Jonathan Mitchley, Joseph Tzanopoulos and Tamsin Cooper, was also built around the development and analysis of scenarios. The project, funded by the European Commission, addressed the biodiversity consequences of scenarios of land-use change in mountain areas within six countries: France, Greece, Norway, Slovak Republic, Switzerland and the UK. As in Sweden, many of these mountain areas are experiencing major demographic changes, and there are conflicts between different groups of stakeholders. However, in contrast to northern Sweden, where agriculture has always been limited because of the high latitude, agricultural activities remain important in maintaining landscape and biological diversity in the six study areas. In recent decades, national and European policies, rather than economic markets, have increasingly become the driving forces of change. The BioScene project developed four scenarios, one 'business as usual', and three which encapsulated the ideas of neoliberalism, biodiversity management and natural ecological processes. Based on these scenarios, developed by a team principally consisting of ecological and socioeconomic scientists, visualizations were used to explore how landscapes and biodiversity might change and how these changes might be accepted by local stakeholders. The outcomes were further integrated and evaluated using sustainability assessment.

The five other case studies are from the relatively sparsely populated mountains of North America and Australia, where landholdings are generally much larger than in Africa or Europe, with significant areas in government ownership. The first three chapters address initiatives of various scales in the Rocky Mountains of Canada; the first of these, by Michael Quinn, Guy Greenaway and Danah Duke, also covers part of northern US, overlapping the area addressed in the grizzly bear case study presented by Mattson and Merrill in their earlier chapter. The chapter considers cumulative effects assessment in this 'Crown of the Continent Ecosystem' (CCE); a process that began with a meeting of representatives of more than 20 government agencies from across the region, on both sides of the international border. The meeting led to the creation of the Crown Managers' Partnership. However, five years later – for reasons explored in the chapter – while the participating agencies still recognize the importance of regional collaboration and integrated management, the goal of using comprehensive computer modelling to explore and quantify the effects of diverse land uses and natural disturbance regimes across the region has not been realized.

The Muskwa-Kechika Management Area (M-KMA) is at the other end of the Canadian Rocky Mountains, in northeastern British Columbia (BC). This is one of the largest unroaded wilderness areas in North America: 6.4 million hectares, larger than many European countries. As with the CCE, a large number of government agencies are involved in managing the diverse resources of the area, including oil, gas, minerals, forests and wildlife. A further complexity is added by the fact that a large proportion of the area is subject to land claims by indigenous people. When the BC government established the M-KMA in 1998, they created a number of administrative, advisory and funding structures, with a strong emphasis on coordinated and comprehensive planning. However, as Paul Mitchell-Banks relates, there have been many challenges to implementing this. While there have been some successes, there have also been failures, stemming particularly from the limited resources and lack of integrated working in the government agencies, constraints from industry and political pressure.

The next two chapters both address ecological restoration in national parks: in Banff in the Canadian Rocky Mountains, and along powerline corridors in the Australian Alps. In Banff National Park, as described by Clifford White and William Fisher, ecological restoration is a relatively new emphasis of park management, as outlined in the 1997 Park Management Plan. This derived from an extensive collaborative process involving diverse stakeholders who developed a widely accepted vision and goals, and continue to be involved in shared decision-making. The adaptive implementation of the plan, both within and adjacent to the park, has been supported by extensive research addressing themes within archaeology, history, fire frequency, wildlife and vegetation ecology, and human dimensions. The experience and knowledge gained through these processes have also contributed to the development and implementation of the management plans for other Canadian national parks and of ecosystem restoration programmes elsewhere in the Rocky Mountains. The complementary chapter written by Stuart Johnston and Roger Good addresses a more specific issue: the integrated restoration and rehabilitation of powerline corridors in the national parks of the Australian Alps. This issue brings together environmental, economic and societal concerns, and the project has involved a wide range of stakeholders. As in Banff, on-the-ground actions have been supported by extensive research. The work has resulted in the successful rehabilitation of powerline corridors, with increasing levels of biodiversity in stable recreated landscapes and ecosystems. Much of the knowledge gained is being disseminated and used more widely in Australia.

The final case study presents a major research programme on climate change, a 'wicked problem' that is likely to be a major force of change throughout the world's mountains in the 21st century (Beniston, 2005) and thus an increasing focus of research attention (for example, Björnsen et al, 2005; Björnsen Gurung, 2006). The Climate Landscape Interactions – Mountain Ecosystem Transect (CLIMET) described by Daniel Fagre, David Peterson and Donald McKenzie, is a major interdisciplinary project to investigate and project the influence of climatic variability and change in three mountain regions, each with a national park at its core, in northwestern US. The project has involved the compilation of existing data, the collection of new data, and the use of computer models to provide information for resource managers across the region. The research has revealed significant changes in glaciers, snow pack and the frequency of natural hazards, and complex responses of forest ecosystems – and has projected possible future trends that are already being considered in management planning. Visual outputs from the modelling work have proven to be key to creating greater understanding of the likely impacts of climate change to park managers, decision-makers, the public and the media. The project has also led to the establishment of a research consortium that incorporates almost all of the mountain systems of western US.

The final chapter brings together many of the lessons learned from both the good and bad experience, and the successes and failures, presented in both the thematic chapters and the case studies. The principles of integrated research and management may not be so different in mountain areas than in other regions. However, for the many reasons noted above, the maintenance of functioning mountain ecosystems – whether 'natural', managed or created and maintained by human action, and including both their human and non-human species – is vital to the well-being of current and future human generations on Earth. Consequently, it will be particularly important in this uncertain century and beyond that there is long-term and effective cooperation between all stakeholders in mountain areas (many of whom live elsewhere) in order to understand and address 'wicked problems'. I hope that this book provides experiences and principles that will be useful and widely applicable in contributing to this goal.

References

Anon (2004) *Programme of Work on Mountain Biodiversity*, Montreal, Secretariat of the Convention on Biological Diversity

Balsiger, P. W. (2004) 'Supradisciplinary research: History, objectives and rationale', *Futures*, vol 36, pp407–421

Bandyopadhyay, J., Rodda, J. D., Kattelmann, R., Kundzewicz, Z. W. and Kraemer, D. (1997) 'Highland waters: A resource of global significance', in Messerli, B. and Ives, J. (eds) *Mountains of the World: A Global Priority*, New York, Parthenon

Beniston, M. (2005) 'The risks associated with climatic change in mountain regions', in Huber, U. M., Bugmann, H. K. M. and Reasoner, M. A. (eds) *Global Change and Mountain Regions: An Overview of Current Knowledge*, Dordrecht, Springer

Berkes, F., Colding, J. and Folke, C. (eds) (2003) *Navigating Social-Ecological Systems: Building Resilience for Complexity and Change*, Cambridge, Cambridge University Press

Bernbaum, E. (1997) *Sacred Mountains of the World*, San Francisco, Sierra Club Books

Björnsen, A., Huber, U., Reasoner, M., Messerli, B. and Bugmann, H. (2005) 'Future research directions', in Huber, U. M., Bugmann, H. K. M. and Reasoner, M. A. (eds) *Global Change and Mountain Regions: An Overview of Current Knowledge*, Dordrecht, Springer

Björnsen Gurung, A. (2006) *Global Change and Mountain Regions (GLOCHAMORE) Research Strategy*, Zurich, Mountain Research Initiative

Blumenthal, D. and Jannink, J-L. (2000) 'A classification of collaborative management methods', *Conservation Ecology*, vol 4, no 2, p13

Braun, T. and Schubert, A. (2003) 'A quantitative view on the coming of age of inter-disciplinarity in the sciences 1980–1999', *Scientometrics*, vol 58, pp183–189

Bruce, A., Lyall, C., Tait, J. and Williams, R. (2004) 'Interdisciplinary integration in Europe: The case of the Fifth Framework Programme', *Futures*, vol 36, pp457–470

Brunner, R. D. and Clark, T. W. (1997) 'A practice-based approach to ecosystem management', *Conservation Biology*, vol 11, pp48–58

CEM (Commission on Ecosystem Management) (2006) *Biodiversity and Livelihoods: Where the Ecosystem Approach Can Take Us*, Gland, IUCN

Chambers, R. (1994) 'The origins and practice of participatory rural appraisal', *World Development*, vol 22, pp953–969

Checkland, P. and Scholes, J. (1990) *Soft Systems Method in Action*, New York, Wiley

Child, B. (ed.) (2004) *Parks in Transition: Biodiversity, Rural development and the Bottom Line*, London, Earthscan

Convention on Biological Diversity (2000) *Principles of the Ecosystem Approach*, www.biodiv.org/programmes/cross-cutting/ecosystem/principles.asp accessed November 2006

Conway, G. R. and Barbier, E. B. (1990) *After the Green Revolution: Sustainable Agriculture for Development*, London, Earthscan

COSEPUP (Committee on Science, Engineering and Public Policy) (2004) *Facilitating Interdisciplinary Research*, Washington DC, National Academies Press

Dogan, M. and Pahre, R. (1990) *Creative Marginality: Innovation at the Intersections of Social Sciences*, Boulder, Westview

Folke, C., Colding, J. and Berkes, F. (2003) 'Synthesis: Building resilience and adaptive capacity in social-ecological systems', in Berkes, F., Colding, J. and Folke, C. (eds) *Navigating Social-Ecological Systems: Building Resilience for Complexity and Change*, Cambridge, Cambridge University Press

Fry, G., Tress, B. and Tress, G. (2005) 'PhD students and integrative research', in Tress, B., Tress, G., Fry, G. and Opdam, P. (eds) *From Landscape Research to Landscape Planning: Aspects of Integration, Education and Application*, Heidelberg, Springer

Gibbons, M., Limoges, C., Nowotny, H., Schwartzman, S., Scott, P. and Trow, M. (1994) *The New Production of Knowledge: The Dynamics of Science and Research in Contemporary Societies*, London, Sage

Godde, P. M., Price, M. F. and Zimmermann, F. M. (eds) (2000) *Tourism and Development in Mountain Regions*, Wallingford, CABI Publishing

Grötzbach, E. and Stadel, C. (1997) 'Mountain peoples and cultures', in Messerli, B. and Ives, J. (eds) *Mountains of the World: A Global Priority*, New York, Parthenon

Grumbine, R. E. (1994) 'What is ecosystem management?', *Conservation Biology*, vol 8, pp27–38

Grumbine, R. E. (1997) 'Reflections on "What is ecosystem management?"', *Conservation Biology*, vol 11, pp41–47

Gunderson, L. H. and Holling, C. S. (eds) (2002) *Panarchy: Understanding Transformations in Human and Natural Systems*, Washington DC, Island Press

Häberli, R., Bill, A., Grossenbacher-Mansuy, W., Thompson Klein, J., Scholz, R.W. and Welti, M. (2001) 'Synthesis', in Thompson Klein, J., Grossenbacher-Mansuy, W., Häberli, R., Bill, A., Scholz, R. W. and Welti, M. (eds) *Transdisciplinarity: Joint Problem-Solving Among Science, Technology and Society: An Effective Way for Managing Complexity*, Basel, Birkhäuser

Hamilton, L. and Bruijnzeel, L. A. (1997) 'Mountain watersheds – integrating water, soils, gravity, vegetation and people', in Messerli, B. and Ives, J. (eds) *Mountains of the World: A Global Priority*, New York, Parthenon

Harmon, D. and Worboys, G. L. (eds) (2004) *Managing Mountain Protected Areas: Challenges and Responses for the 21st Century*, Colledara, Andromeda Editrice

Haynes, R. W., Szaro, R. C. and Dykstra, D. P. (2005) 'Balancing conflicting values: Ecosystem solutions in the Pacific Northwest of the United States and Canada', in Sayer, J. A. and Maginnis, S. (eds) *Forests in Landscapes: Ecosystem Approaches to Sustainability*, London, Earthscan

Himamowa, B. (1974) The *Obergurgl Model: A Microcosm of Economic Growth in Relation to Limited Ecological Resources*, Laxenburg, International Institute for Applied Ecosystem Analysis

Holling, C. S. (ed.) (1978) *Adaptive Environmental Assessment and Management*, Chichester, Wiley

Huddleston, B., Ataman, E., de Salvo, P., Zanetti, M., Bloise, M., Bel, J., Franceschini G. and Fe d'Ostiani, L. (2003) *Towards a GIS-based Analysis of Mountain Environments and Populations*, Rome, FAO

Hurni, H. (2003) 'The Bishkek Mountain Platform', *Mountain Research and Development*, vol 23, pp86–89

Hurni, H. and Wiesmann, U. (2004) 'Towards transdisciplinarity in sustainability-oriented research for development', in Hurni, H., Wiesmann, U. and Schertenleib, R. (eds) *Research for Mitigating Syndromes of Global Change: A Transdisciplinary Appraisal of Selected Regions of the World to Prepare Development-Oriented Research Partnerships*, Berne, NCCR North-South

Hurni, H., Wiesmann, U. and Schertenleib, R. (eds) (2004) *Research for Mitigating Syndromes of Global Change: A Transdisciplinary Appraisal of Selected Regions of the World to Prepare Development-Oriented Research Partnerships*, Berne, NCCR North-South

Ives, J. D., Messerli, B. and Rhoades, R. E. (1997) 'Agenda for sustainable mountain development', in Messerli, B. and Ives, J. (eds) *Mountains of the World: A Global Priority*, New York, Parthenon

Janssen, M. A. (2002) 'A future of surprises', in Gunderson, L. H. and Holling, C. S. (eds) *Panarchy: Understanding Transformations in Human and Natural Systems*, Washington DC, Island Press

Jeník, J. (1997) 'The diversity of mountain life', in Messerli, B. and Ives, J. (eds) *Mountains of the World: A Global Priority*, New York, Parthenon

Kapos, V., Rhind, J., Edwards, M., Price, M. F. and Ravilious, C. (2000) 'Developing a map of the world's mountain forests', in Price, M. F. and Butt, N. (eds) *Forests in Sustainable Mountain Development: A State-of-Knowledge Report for 2000*, Wallingford, CAB International

Kennedy, J. J. and Koch N. E. (2004) 'Viewing and managing natural resources as human-ecosystem relationships' *Forest Policy and Economics*, vol 6, pp497–504

Körner, C. and Ohsawa, M. et al (2005) 'Mountain systems', in *Millennium Ecosystem Assessment, 2005. Current State and Trends: Findings of the Condition and Trends Working Group. Ecosystems and Human Well-being, Vol.1*, Washington DC, Island Press

Körner, C. and Spehn, E. M. (eds) (2002) *Mountain Biodiversity: A Global Assessment*, New York and London, Parthenon

Lawrence, R. J. and Després, C. (2004) 'Introduction: Futures of transdisciplinarity', *Futures*, vol 36, pp397–405

Lowe, P. and Phillipson, J. (2006) 'Reflexive interdisciplinary research: The making of a research programme on the rural economy and land use', *Journal of Agricultural Economics*, vol 57, pp165–184

Ludwig, D., Mangel, M. and Haddad, B. (2001) 'Uncertainty, resource exploitation, and conservation', *Annual Review of Ecology and Systematics*, vol 32, pp481–517

Luukkonen, T., Nedeva, M. and Barré, R. (2006) 'Understanding the dynamics of networks of excellence', *Science and Public Policy*, vol 33, pp239–252

Mackey, J. L. (2002) 'Rules are not the way to do interdisciplinarity: A response to Szostak', *Issues in Integrative Studies*, vol 20, pp123–129

McCracken, J. A., Pretty, J. N. and Conway, G. R. (1988) *An Introduction to Rapid Rural Appraisal for Rural Development*, London, International Institute for Environment and Development

Messerli, B. and Ives, J. D. (eds) (1997) *Mountains of the World: A Global Priority*, New York, Parthenon

Meybeck, M., Green, P. and Vörösmarty, C. (2001) 'A new typology for mountains and other relief classes: An application to global continental water resources and population distribution', *Mountain Research and Development,* vol 21, pp34–45

Mountain Agenda (1998) *Mountains of the World: Water Towers for the 21st Century*, Berne, Institute of Geography

Mountain Agenda (2001) *Mountains of the World: Mountains, Energy, and Transport*, Berne, Centre for Development and Environment

Nicolescu, B. (2003) 'Definition of transdisciplinarity', www.interdisciplines.org

Norton, B. G. (2005) *Sustainability: A Philosophy of Adaptive Ecosystem Management*, Chicago, University of Chicago Press

Nowotny, H. (2003) 'The potential of transdisciplinarity', www.interdisciplines.org

O'Riordan, T. and Stoll-Kleemann, S. (eds) (2002) *Biodiversity, Sustainability and Human Communities: Protecting beyond the Protected*, Cambridge, Cambridge University Press

Pohl, C. (2005) 'Transdisciplinary collaboration in environmental research', *Futures*, vol 37, pp1159–1158

Pound, B., Snapp, S., McDougall, C. and Braun, A. (eds) (2003) *Managing Natural Resources for Sustainable Livelihoods: Linking Science and Participation*, London, Earthscan

Prato, T. and Fagre, D. (2005) *National Parks and Protected Areas: Approaches for Balancing Social, Economic and Ecological Values*, Oxford, Blackwell

Price, M. F. and Butt, N. (eds) (2000) *Forests in Sustainable Mountain Development: A State-of-Knowledge Report for 2000*, Wallingford, CAB International

Price, M. F. and Hofer, T. (2005) 'The International Year of Mountains, 2002: Progress and prospects', in Thompson, D. B. A., Price M. F. and Galbraith, C. A. (eds) *The Mountains of Northern Europe: Conservation, Management, People and Nature*, Edinburgh, The Stationery Office

Price, M. F. and Thompson, M. (1997) 'The complex life: Human land uses in mountain ecosystems', *Global Ecology and Biogeography Letters*, vol 6, pp77–90

Price, M. F., Moss, L. A. G. and Williams, P. W. (1997) 'Tourism and amenity migration', in Messerli, B. and Ives, J. (eds) *Mountains of the World: A Global Priority*, New York, Parthenon

Rhoades, R. E. (1997) *Pathways Towards a Sustainable Mountain Agriculture for the 21st Century: The Hindu-Kush Experience*, Kathmandu, International Centre for Integrated Mountain Development

Rittel, H. and Webber, M. (1973) 'Dilemmas in a general theory of planning', *Policy Sciences*, vol 4, pp155–169

Royal Swedish Academy of Sciences (2002) *The Abisko Agenda: Research for Mountain Area Development*, Ambio Special Report, Stockholm, Royal Swedish Academy of Sciences

Sayer, J. and Campbell, B. (2004) *The Science of Sustainable Development: Local Livelihoods and the Global Environment*, Cambridge, Cambridge University Press

Schweizer, P. and Preiser, K. (1997) 'Energy resources for remote highland areas', in Messerli, B. and Ives, J. (eds) *Mountains of the World: A Global Priority*, New York, Parthenon

Secretariat of the Convention on Biological Diversity (2000) *Conference of the Parties Decisions. Decision V/6 Ecosystem Approach*, Geneva, United Nations Environment Programme

Shepherd, G. (2004) *The Ecosystem Approach: Five Steps to Implementation*, Gland and Cambridge, IUCN

Smith, R. D. and Maltby, E. (2003) *Using the Ecosystem Approach to Implement the Convention on Biological Diversity*, Gland and Cambridge, IUCN

Spehn, E. M., Liberman, M. and Körner, C. (eds) (2006) *Land Use Change and Mountain Biodiversity*, Boca Raton, CRC Press

Sperber, D. (2003) 'Why rethink interdisciplinarity', www.interdisciplines.org

Stepp, J. R., Castaneda, H. and Cervone, S. (2005) 'Mountains and biocultural diversity', *Mountain Research and Development*, vol 25, pp223–227

Stewart, R. E., Walters, L. C., Balint, P. J. and Desai, A. (2004) *Managing Wicked Environmental Problems*, Report to Jeff Blackwell, Regional Forester, USDA Forest Service, Pacific Southwest Region, Fairfax, George Mason University

Stolton, S. and Dudley, N. (eds) (1999) *Partnerships for Protection: New Strategies for Planning and Management for Protected Areas*, London, Earthscan

Stone, P. B. (ed.) (1992) *The State of the World's Mountains*, London, Zed Books

Stone, P. B. (2002) 'The fight for mountain environments', *Alpine Journal*, vol 107, pp117–131

Szostak, R. (2002) 'How to do interdisciplinarity: Integrating the debate', *Issues in Integrative Studies*, vol 20, pp103–122

Taylor, L. and Ryall, A. (eds) (2004) *Interdisciplinary Research and Management in Mountain Areas*, Banff, The Banff Centre

Thompson Klein, J. (1990) *Interdisciplinarity: History, Theory and Practice*, Detroit, Wayne State University Press

Thompson Klein, J. (2004) 'Prospects for transdisciplinarity', *Futures*, vol 36, pp515–526

Thompson Klein, J., Grossenbacher-Mansuy, W., Häberli, R., Bill, A., Scholz, R. W. and Welti, M. (eds) (2001) *Transdisciplinarity: Joint Problem-Solving Among*

Science, Technology and Society: An Effective Way for Managing Complexity, Basel, Birkhäuser

Tress, B., Tress, G. and Fry, G. (2005) 'Integrative studies on rural landscapes: Policy expectations and research practices', *Landscape and Urban Planning*, vol 70, pp177–191

United Nations (2002) *Report of the World Summit on Sustainable Development*, New York, United Nations

Viviroli, D., Weingartner, R. and Messerli, B. (2003) 'Assessing the hydrological significance of the world's mountains', *Mountain Research and Development*, vol 23, pp32–40

Walters, C. (1986) *Adaptive Management of Renewable Resources*, New York, Macmillan

Weingart, P. and Stehr, M. (1999) *Practising Interdisciplinarity*, Toronto, University of Toronto Press

Wu, J. (2006) 'Landscape ecology, cross-disciplinarity, and sustainability science', *Landscape Ecology*, vol 21, pp1–4

Zimmerer, K. S. (ed.) (2006) *Globalization and New Geographies of Conservation*, Chicago, University of Chicago Press

Zorn, P., Stephenson, W. and Grigoriev, P. (2001) 'An ecosystem management program and assessment process for Ontario National Parks', *Conservation Biology*, vol 15, pp353–362

Zube, E. (1982) 'Increasing the effective participation of social scientists in environmental research and planning', *International Social Science Journal*, vol 34, pp481–492

2

From Local Projects in the Alps to Global Change Programmes in Mountain Areas: The Development of Interdisciplinarity and Transdisciplinarity in the Last 25 Years

Bruno Messerli and Paul Messerli

Introduction: Interdisciplinary mountain research 200 years ago

Alexander von Humboldt conducted his most innovative fieldwork on Chimborazo (6267m), in Ecuador's volcano valley in 1802, exactly 200 years before the International Year of Mountains in 2002. He did not restrict his studies to simple observations of different altitudinal belts; he also included climate, plant life, geology and cultivated crops, and he measured several physical parameters (see Box 2.1) in order to understand the relationship and interactions between different elements and between natural processes and human activities.

During his journeys through the tropical Americas (1799–1804), von Humboldt amassed volumes of primary data on flora and fauna, geomagnetism and volcanism, ocean currents, archaeological treasures and cultural features: the raw material for all his subsequent lectures and writing. In 1799 he wrote to a good friend (Buttimer, 2004; De Terra, 1955):

> *I shall collect plants and fossils, and make astronomical observations with the best of the instruments. Yet this is not the main purpose of my journey. I shall endeavour to discover how nature's forces act upon one another and in what manner the geographic environment exerts its influence on animals and plants. In short, I must find out about the harmony in nature.*

Box 2.1 Tableau Physiques des Régions Equitoriales

Observed and measured parameters in the equatorial mountains of South America between 10°N and 10°S during the years 1799–1803:

- Vegetation
- Animal life
- Geological formations
- Cultivated crops
- Air temperature
- Perennial snow line
- Chemical composition of the atmosphere
- Barometric pressure
- Gravitational decrease
- Colour intensity of the azure sky
- Degree at which water boils at different elevations

Humbolt wrote: 'I have tried to combine the entire range of phenomena observed in tropical regions, from sea level to the highest peaks of the Andes'.

Source: Humboldt (1805)

Box 2.1 shows the parameters that von Humboldt used for his studies on Chimborazo and in other mountain systems of the tropical Americas. Mountains had a special appeal for von Humboldt. Mountains themselves, their profiles, vegetation carpets and their inhabitants' ways of life played a vital role in the evolution of his world-view. Mountains posed challenges for not only physical endurance and unprecedented scientific analysis, they also inspired innovative ways of presenting research results in graphic forms (Buttimer, 2004). Figure 2.1 shows a summary – unique for this time – of all the local, place-based findings in the Andes, Tenerife, Himalayas, Alps, Pyrenees and even Lapland in an integrated and quite worldwide comparison. Looking at this fascinating figure, we should remember that behind this most impressive result was innovative thinking, which we must designate as highly interdisciplinary. This may be shown in the three following sentences (Humboldt, 1845–1862):

- *In these immense chains of cause and effect, nothing can be regarded in isolation.*
- *The overall equilibrium which exists throughout major disturbations is the result of an infinite range of mechanical forces and chemical reactions, all of which balance each other.*

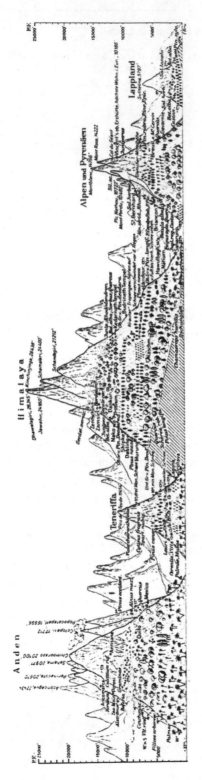

Figure 2.1 Comparison of dominant vegetation belts, with temperature data, in different mountain systems of the world

Source: Humboldt (1845–1862)

- *While each series of facts needs to be studied separately in order to discover its own rules of order, the general study of nature demands that all knowledges ... be then combined.*

From this brilliant and ingenious way of thinking, 200 years ago, that is truly interdisciplinary, we could learn a great deal for present and future interdisciplinary projects in the mountains of the world.

The Swiss MAB Mountain Programme (1979–1987)

In the 1970s and 1980s, the worldwide Man and Biosphere (MAB) Programme of UNESCO had a stimulating effect on research in the Alps and on cooperation between Alpine countries. The Swiss MAB Programme on 'socio-economic development and ecological carrying capacity in mountain regions' began in 1979 in close collaboration with the Austrian Obergurgl project (Patzelt, 1987). Financed by the Swiss Science Foundation, about 40 research studies involving very different disciplines were initiated in four test areas of the Swiss Alps: Grindelwald, Aletsch, Pays d' Enhaut and Davos. These four regions were very different from each other in their natural, economic, cultural and political conditions. However, all were located in the Alps – in the Swiss periphery – away from the powerful political and economic centres. Such mountain regions are often controlled by external forces that use the regions' resources and influence their political decision-making processes; a common situation for mountain regions worldwide (Messerli, 1983). The main aim of the Swiss MAB Programme was to understand the driving forces, the transformation of these communities, and the relationships between inhabitants and their environment; to identify desirable and undesirable structures and processes; and finally to answer the question of how long-term well-balanced development could be achieved at a time when 'sustainability' was still an unknown concept.

Before discussing new methods and models in general, and interdisciplinary approaches in particular, we should consider the dramatic change of Alpine mountain communities from the second half of the 19th to the second half of the 20th centuries; without understanding the past we shall never understand the future! For Obergurgl, for instance, Moser (1975) was able to reconstruct the situation around 1860 in this mountain village. Obergurgl could be represented as a closed system (see Figure 2.2) with a strong relationship between the inhabitants and the environment with its natural resources. Over generations, the farmers had learned the meaning of 'sustainability' because the system was virtually closed and resources were limited. If the population increased to a level that exceeded the capability of the area to support them, some had to leave. Therefore a strong social order was developed to maintain the population number in balance with the ability to produce the necessary food. In these circumstances, a modelling approach would have

Closed System

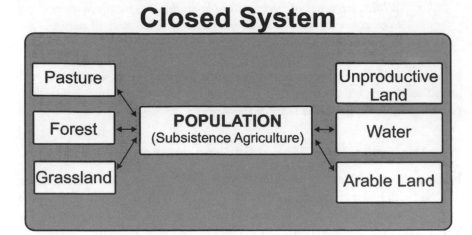

Open System

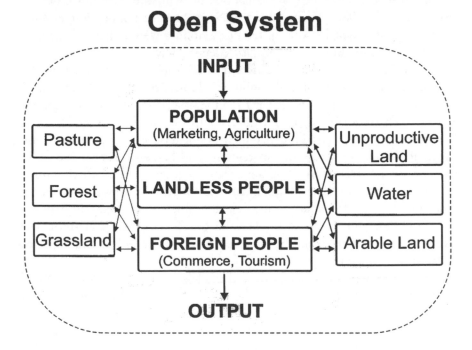

Figure 2.2 Obergurgl, Austria: Change from a more-or-less closed system around 1860 to an open system in recent years

Source: Based on Moser (1975), especially the open system

been relatively simple: the areas under grassland, forest and arable land required to support a specific number of people could be calculated. On this basis, it would be possible to develop a number of scenarios (for example, climatic deterioration or population growth) and draw quite reasonable conclusions.

These immediate challenges and response relationships between a human population and their environment no longer exist due to the opening up of the system through increasing encroachments of external processes and forces. Figure 2.2 also presents a modification of Moser's open system so that it can represent a local community anywhere in the world. There are now inputs and outputs of people, capital, energy, technology, information, goods and services in many different forms. The relationships between the elements of the system are very complex, sometimes invisible, sometimes disturbed, and often changing. Even the areas perceived as unproductive under traditional use have become important for tourist infrastructure, dam construction and aesthetic qualities. To model such a system, not knowing whether some key elements are external (controlled from an outside political or economic centre) or internal, is much more difficult – and without a qualified interdisciplinary team it is quite impossible to develop good solutions and longer-term strategies.

In view of these problems, an elementary scheme was developed early in the Swiss MAB Programme, in which the most important structures and processes in a specific test area are placed in a simplified design (see Figure 2.3) (Messerli and Messerli, 1978). It can be used to indicate the fundamental processes, including the influence of external factors (relationship 1). The socio-economic system can be derived from several subsystems depending upon local conditions (for example, economic, political, demographic, cultural). All the activities emanating from this socio-economic system influence not only the type and intensity of land use, but also the ecosystem's quality and services (relationship 2). If the intensity of use can be absorbed by the natural conditions, there will be a normal feedback from the land-use to the socio-economic systems, for example in the form of a good harvest (relationship 3). If it cannot be absorbed, negative effects will damage the natural system (relationship 4), and this could have negative impacts on the land-use (relationship 5) and even the socio-economic systems. This preliminary model was only a first attempt to clarify which flow paths (relationships 1–5) could and should be investigated. It follows that all the relationships need to be examined through a highly interdisciplinary approach – and also that the land-use subsystem, which forms the link between the demands of society and the capacity of nature to accommodate those demands, is the variable to be optimized in the human–environment system (Ives and Messerli, 1990; Messerli, 1986).

In the four Swiss MAB test areas, new methods and models were developed and applied, taking into account the interactions between the natural and socio-economic system. This modelling approach also forced the researchers to develop different data structures and data banks for interdisciplinary projects. Moreover, simulation models bacame an important instrument in discussions with local authorities. In several meetings, the demonstration of simulation results initiated stimulating dialogues between scientists and local representatives. Moreover, confidence in scientific results created the goodwill to discuss

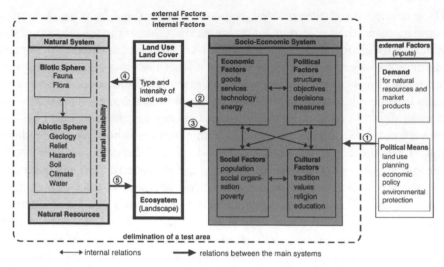

Figure 2.3 Schematic representation of a regional economic–ecological system: Swiss MAB Programme

Source: Messerli and Messerli (1978)

unprecedented solutions and actions. In retrospect, the programme was a continuous balancing act between holistic and reductionist approaches and between scientific knowledge and experience, to guarantee a permanent inter-action between science and society (Messerli, 1986, 1989).

Figure 2.4 shows the final application-oriented hierarchical MAB model for the formulation of long-term strategies. The different levels of the model represent different stages of knowledge and integration. However, interactions and decision-making depend very strongly on positive or negative external factors ('top-down') and on the active or passive behaviour of the local popu-lation ('bottom-up').

To illustrate the Swiss MAB Programme, two of the four test regions are briefly presented below, each with a focus on a certain topic. The case study of Aletsch in the central Alps shows the dramatic impact of economic forces, especially tourism, on a traditional mountain population and its land-use system in the second half of the 20th century. The case study of Grindelwald, a historical centre of mountain tourism, demonstrates the comprehensive application of the final MAB model (see Figure 2.4) and documents the process of implementing the results from 1986 until today.

The Aletsch Region

The Aletsch Region (Messerli, 1983; Messerli et al, 2003) lies in the upper part of the Rhone valley in the central Alps. It is bounded by the longest glacier

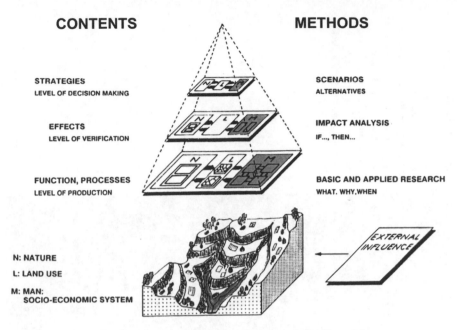

CONTENTS

STRATEGIES
LEVEL OF DECISION MAKING

EFFECTS
LEVEL OF VERIFICATION

FUNCTION, PROCESSES
LEVEL OF PRODUCTION

N: NATURE

L: LAND USE

M: MAN:
SOCIO-ECONOMIC SYSTEM

METHODS

SCENARIOS
ALTERNATIVES

IMPACT ANALYSIS
IF..., THEN...

BASIC AND APPLIED RESEARCH
WHAT, WHY, WHEN

Figure 2.4 Further development of the Swiss MAB model

Note: The model develops through different levels of contents and methods in order to reach a satisfactory base for scenarios, strategies and decision-making

Source: Messerli (1986)

and the largest glaciated area of the Alps, and at its northern edge includes the summits of the Jungfrau (4089m) and Mönch (4099m) on the border of the Bernese Oberland or the northern Alps. The area includes several communities and, during the International Year of Mountains in 2002, was designated the first Natural World Heritage Site in the Alps (see Figure 2.5). The original land use of the inner-alpine region was determined by the need for self-sufficiency. In particular, relief and climate determined the type of land use (see Figure 2.6) as:

- the area with permanent settlements including the surrounding fields (*Heimgüter*);
- the pre-alpine region with seasonally inhabited settlements (*Maiensäss*);
- the alpine region with summer pastures (*Alps*).

The intensity of land use decreased from 'valley to alp', that is, from ploughed fields at the lowest level, to permanent meadows and pastures at the middle level, and to summer pastures in the highest level. Thus, the system was adapted to the increasing sensitivity of the ecosystems with increasing altitude. Analysis of climate and vegetation shows a decreasing vegetation period from

Figure 2.5 The Aletsch glacier

Note: The Aletsch glacier is the longest glacier in the Alps (23km) and is in the centre, with the Rhone valley in the foreground and the highest peaks (above 4000m) of the Bernese Oberland in the background (from a tourist leaflet). The slope between the Rhone valley and the Aletsch glacier valley is addressed in Figure 2.6.

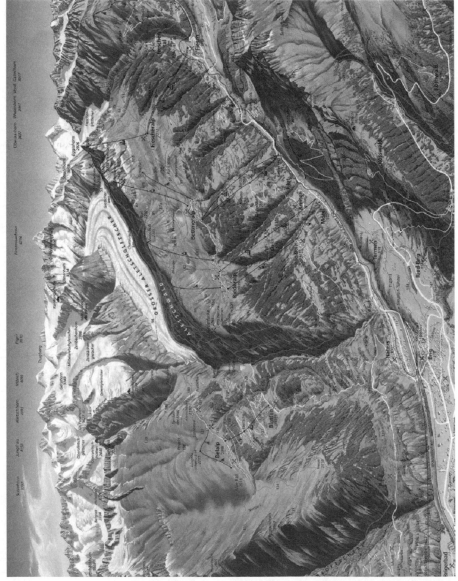

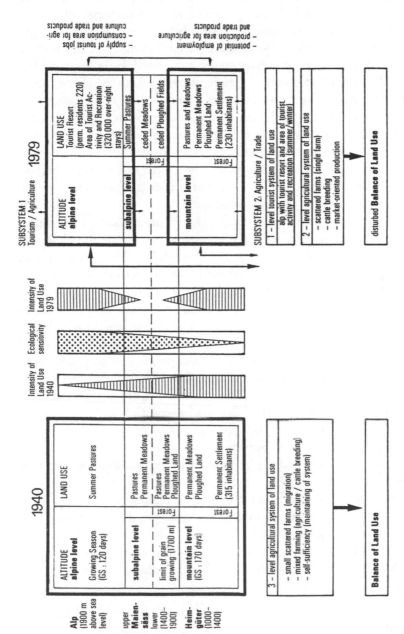

Figure 2.6 Test area Aletsch of the Swiss MAB Programme

Note: The figure shows the consequences of the changes from a relatively closed system until the end of the second world war to an open system in the second half of the 20th century.

Source: Messerli (1983)

more than 200 days at the original permanent settlements to 100 days and less at the summer pastures above the timberline (Messerli et al, 2003). These data may show the perfect adaptation of the population and its land-use system to the natural conditions, without knowing the definition of sustainability.

In the 1950s, the region was made accessible by cable car, and this brought about the displacement of the self-sufficient closed system. Most of the rural population moved to the newly created, more attractive jobs in tourism. The subsequent reduction in the number of agricultural workers led to the break-down of the three-level agricultural system and to the abandonment of disadvantageously situated agricultural land. Moreover, people now employed in tourism changed their domicile from the original permanent settlement to the newly opened alpine tourist settlement. Today, instead of the single origi-nal economic unit, there are two subsystems: the tourist system at the highest level, with a tendency towards expansion; and the agriculture system at the lower levels, with a tendency towards reduction.

The consequences of this excessively fast change are obvious. The original balance can no longer be maintained; the ecologically sensitive alpine belt is used most intensively by growing settlements and expanding tourist infrastructure, including transport systems, paths and ski runs. The protective and productive functions of the subalpine forest belt are endangered by improper or inadequate management. This belt, together with the abandoned fields, forms a zone of potential instability, with a danger of landslides and avalanches. In addition to these ecological consequences, there are changes in the economic and social spheres. Differences in income between the rural population and the people involved in the tourist industry influence the political balance and change values, expectations and patterns of behaviour. These changes cannot be discussed here in detail, but can be summarized in the following five points:

1 divergence of agriculture and tourism;
2 'critical mass of tourism' and the importance of the qualification structure of the labour market;
3 spreading of economic risk in mono-structured regions;
4 maintenance of self-regulating capacity depending on the behaviour of the local population and authorities;
5 'cultural lag' – the gap between a new socio-economic reality and tradi-tional value systems, which is augmented as the rate of economic change increases.

Grindelwald

Grindelwald (Wiesmann, 2001) lies immediately below the highest summits of the Bernese Oberland, and this spectacular landscape – combined with the short transport distance from Berne – was the basis for tourism, which began in the 19th century and was highly developed even before the First World War.

Figure 2.7 The Eiger with its famous north wall is the dominant mountain on this tourist leaflet

Note: On the left is the valley of Grindelwald and on the right the valley of Lauterbrunnen. With the exceptions of the railways to Grindelwald, Lauterbrunnen and Jungfraujoch, all the other transport systems, especially for winter sport, were built in the second half of the 20th century.

The official inauguration of the railway to the Jungfraujoch (3454m) in 1912 may be regarded as a symbol for the development boom of this period. Other astonishing projects to reach the mountain tops of the Eiger and Jungfrau failed with the beginning of the First World War. Based on this long experience with tourism and with economic forces outside agriculture, the MAB research project was very welcome and was immediately of interest to the local authorities and local population because the fast expansion of tourist infrastructure, especially for winter sports, had had a profound impact on the landscape (see Figure 2.7). Without going into the detail of all the different disciplinary and interdisciplinary research projects, we concentrate on the long-term participatory process between the local population and the researchers to show the value of interdisciplinary results and the possibilities of implementation through an intense interaction between science and society.

1976 to 1986
Interdisciplinary research was carried out in the region of Grindelwald by a team of researchers from 12 different institutions, covering a broad range of natural and social sciences. The synthesis of the MAB Programme Grindelwald illustrated that a subtle balance between tourism, agriculture and the environment had developed over time. However, this positive situation was building on socio-economic structures that tend to destroy the subtle balance in the long term.

1984 to 1987
Based on the results of the MAB Programme, the community of Grindelwald initiated a participatory process in order to formulate certain policy guidelines. This process involved all inhabitants and more than 70 local civil society institutions. Members of the MAB research team acted as resource persons. This process resulted in policy guidelines – 'Grindelwald 2000' (Wiesmann, 2001) – that explicitly formulated long-term goals covering the environmental, socio-cultural and economic dimensions of sustainability. The goals were addressed to all relevant stakeholders. 'Grindelwald 2000' can be seen as the first concretely realized Local Agenda 21 in the Alps.

1986 to 2000
The community and institutions of Grindelwald developed and implemented measures compatible with the policy guidelines. A key concern was to break the structural forces that endanger the positive balance between tourism, agriculture and the environment. Of utmost importance were measures in the following four fields: controlling investments in the transport sector; steering and channelling building investments without endangering the labour market; reviewing the village and landscape plans; and diversifying the elements of the tourist supply.

1999 to 2001

The community commissioned the former MAB research team to scientifically monitor the impacts of 'Grindelwald 2000'. At the same time, the community politically evaluated the effects of the guidelines. These assessments disclosed that they had generally been successful and had a broad range of direct and indirect effects. However, the evaluation also led to the conclusion that time had come for a new round of negotiations in the community on the goals of sustainability.

Conclusions from the Swiss MAB Programme

The creation of interdisciplinary research teams in the Swiss MAB Programme was not easy. A number of factors that led to their success can be identified: strong leadership from a multidisciplinary and committed expert group with a high scientific reputation, open to interdisciplinary approaches; excellent programme management; clearly defined research themes and approaches that attracted qualified scientists; and, based on these conditions, adequate funding from the National Science Foundation.

The Swiss MAB Programme led to advances in the theory and practice of human–environment research in other mountain regions of Europe, Russia and Africa (Price, 1995). For the Alps, it was concluded that six conditions must be met in order to achieve consistent development (Brugger et al, 1984):

1   The mountain regions should remain as living space for an active population with a balanced rate of out- and in-migration.
2   Self-reliance in the use of the region's economic potential is necessary.
3   The maintenance of a distinct local landscape is a prerequisite for some economic activities, especially tourism.
4   Since the mountain regions support a unique variety of flora and fauna, biodiversity must be preserved; this is important at both national and international levels.
5   Protection and utilization, 'carrying capacity' and sustainability all require that the indigenous institutions of the mountain regions retain a certain degree of political autonomy.
6   Self-reliant political institutions and practices must be supported by cultural individuality, variety and autonomy. The maintenance and support of local and cultural diversity is therefore of primary importance.

Overall, interdisciplinarity was a prerequisite for a successful Swiss MAB Programme. Yet, in many of its projects, this research programme passed the borderline from interdisciplinarity to transdisciplinarity, as defined and discussed in the following section.

From interdisciplinarity to transdisciplinarity, especially in the developing world since 1986

While fieldwork on highly specialized topics was under way in the mountains of Africa by the late 1960s (Messerli, 1980), integrated projects with natural and social science components started only in the 1980s, reaching an initial culmination with the first conference on African mountains and the foundation of the African Mountain Association in Addis Ababa in 1986 (Messerli and Hurni, 1990). Research projects in the African context – for example on soil erosion, a crucial problem in the mountains and highlands of East Africa – showed very clearly that pure natural science approaches were worthless without understanding and integrating the social, economic and demographic conditions of the local population (Messerli et al, 1988). Moreover, fieldwork in the mountains of tropical and subtropical Africa indicated very quickly that the MAB model, created in and for the Alps, was not satisfactory for mountain areas in poor developing countries with difficult and unstable political conditions, where extreme problems such as land degradation, water scarcity, population growth, poverty and even hunger dominate the entire natural–human system. Key processes determine the pattern, intensity and techniques of land use. If these activities are not adapted to the natural system, they will provoke serious environmental conflicts. Equally, certain limitations are defined by the natural system and, if not in equililbrium with human-induced processes, they damage the land-use and socio-economic systems. In this context, the Alpine model was adapted and changed, and key processes and limiting factors were introduced (see Figure 2.8), which also forced the researchers to set priorities and to start emergency actions. More precisely, the key processes and limitations must be clearly identified and introduced in decision-making processes at all levels, from the political authorities to the single farmer (Messerli et al, 1988). Thus real and place-based problems, involving science and society, became the starting point for integrated and transdisciplinary projects in the mountains of the developing world.

In the industrialized world, transdisciplinary research means addressing concrete problems of society and working out solutions through cooperation between actors and scientists (Thompson Klein et al, 2001). Although transdisciplinarity may be a relatively new word, the concept of addressing concrete problems of society and initiating joint problem-solving involving science, technology and society in order to understand and manage complexity more efficiently has a long tradition. However, in the 1970s, when the concepts for the different UNESCO MAB programmes were developed, this special definition of transdisciplinarity did not yet exist, although the participation of the local population and the political authorities was an important element in all the project guidelines. Through an international conference, based on the results of a 'Priority Programme on the Environment' of the Swiss National Science Foundation, transdisciplinarity has become an important topic in

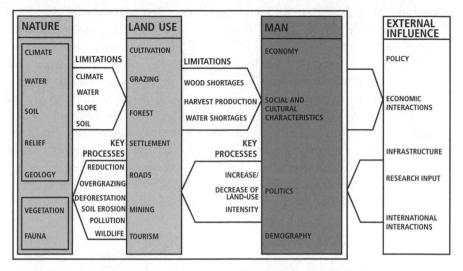

Figure 2.8 The Swiss MAB model adapted to the conditions
of the mountains and highlands in Ethiopia

Note: Key processes and limiting factors determine transdisciplinary and integrated approaches

Source: Hurni in Messerli et al (1988)

recent years and can be described, with some key points, as follows (Thompson Klein et al, 2001):

- The core idea of transdisciplinarity is that different academic disciplines work jointly with practitioners to solve a real problem. It can be applied in great variety of fields.
- Transdisciplinary research is an additional type within the spectrum of research and coexists with traditional monodisciplinary research.
- The science system is the primary knowledge system in society. Transdisciplinarity is a way of increasing its unrealized intellectual potential and its effectiveness.
- Transdisciplinary projects need clear goals and a competent management to facilitate creativity and minimize friction among members of the team.
- Stakeholders must participate from the beginning and be kept interested and active over the entire course of a project.
- Courageous research and development administration is needed to promote transdisciplinarity, not praise interdisciplinarity and still promote disciplinarity.

For the developing world, the transdisciplinary approach became the basis for the further development of 'syndrome mitigation research'. The concept of a syndrome of global change was borrowed from the German Advisory Council on Global Change, which first looked at combinations or clusters of problems of non-sustainable development (Hurni et al, 2004). The Swiss National Centre of Competence in Research (NCCR) North-South revised the German version, which was criticized for having a focus only on negative aspects of development. Therefore, priorities are given to innovative approaches to mitigating syndromes, to participatory and sustainability-oriented research and, also, to normative dimensions of sustainability. This new direction for syndrome mitigation research was defined in a programme entitled, 'Research for Mitigating Syndromes of Global Change. A Transdisciplinary Appraisal of Selected Regions of the World to Prepare Development-Oriented Research Partnerships' (Hurni et al, 2004; Messerli et al, 2005). Three long-term objectives have been defined to achieve the main aim of the programme (Hurni et al, 2004):

1   to promote disciplinary, interdisciplinary and transdisciplinary research focusing on sustainable development ('transdisciplinarity' in this context is understood as an approach based on collaboration with local people that takes account of their rich knowledge);
2   to help develop institutions and train staff in these fields of research, in partner countries and in Switzerland;
3   to support societies and institutions in partner countries in their autonomous efforts to address syndromes of global change over the long term.

The NCCR North-South syndrome mitigation programme is working worldwide in nine regions, seven in developing countries with the two syndrome contexts of 'Mountains' and 'Highland–Lowland Interactions': East Africa, Horn of Africa, Central Asia, South Asia, Mekong region, Central America and Andes. Most interesting is the experiment to compare core problems, evaluated in a careful process by representatives of the local population, national institutions and foreign experts. The analysis of the two syndrome contexts for four regions (East Africa, Horn of Africa, Mekong region, Andes) shows the following preliminary results:

•   Similar core problems identified in the Mountains context:
    –   unequal distribution of power and resources, corruption;
    –   lack of adequate infrastructure and management such as transport, energy and irrigation;
    –   degradation of land, soil and vegetation cover;
    –   degradation of forests and natural habitats.
•   Similar core problems identified in the Highland–Lowland Interactions context:
    –   contradictory policies and weak formal institutions at different levels;
    –   erosion of traditional and/or indigenous institutions;

- – social, cultural and ethnic tensions and insecurity;
- – unfavourable dynamics and imbalances in socio-demographic structures.
- • Similar core problems identified in both the Mountains and the Highland–Lowland Interactions contexts:
  - – governance failures, insufficient empowerment and decentralization;
  - – poverty and insecurity of livelihoods;
  - – inequality of ownership and access to land, natural and common-property resources.

These core problems occur in the same combination in all the analysed regions and can therefore be seen as typical patterns. It is too early to refer to these major patterns of core problems as proven syndromes of global change in mountain areas. First, there is a need for further evidence from ongoing research on the dynamics and impacts of these core problems and their underlying causes. In addition, efforts must be made to understand the interrelations among different core problems, and to compare these problems between all the different mountain regions where the NCCR programme is being implemented (Messerli et al, 2005). Most interesting is the methodological development, from the first idea to change the well-balanced MAB model towards a focus on certain priority problems in the 1980s (from Figure 2.3 to Figure 2.8; see also Funnell and Price, 2003), and to continue from there to the syndrome mitigation approach, which has developed this idea much further with the aim of reaching, through region- and context-specific research, a generalization of core problems as a true contribution to global change research. Thus, disciplinary, interdisciplinary and transdisciplinary approaches could and should each be used not only in the right place for the right question, but together as part of an integrated and focused project.

Global change and mountain research since 1992/1997

UNESCO's MAB Programme covered the main ecosystems on Earth, but was not a global programme in the current sense of understanding global change processes, even if the growing impact of human activities on the environment was the overall common topic for all its different pioneer projects. Only in the second half of the 1980s was the scientific community forced to take new responsibilities by creating new structures and organizations to follow the rapidly growing processes of economic globalization, urbanization and industrialization, with all their consequences for climate and environment. In 1986, the International Council of Scientific Unions (ICSU) established the International Geosphere-Biosphere Programme (IGBP) and in 1996 the International Social Science Council (ISSC) established the International Human Dimensions of Global Environmental Change Programme (IHDP). However, the real breakthrough for mountains as a global topic was the 'Earth

Summit' in Rio de Janeiro in 1992, where Agenda 21 included a mountain chapter entitled, 'Managing Fragile Ecosystems – Mountain Sustainable Development' (Centre for Our Common Future, 1993).

Five years later, in 1997, at a special meeting of the UN General Assembly in New York, a new period for mountain research on a global level began with the support of UNU and UNESCO. This showed that the necessary awareness of the high significance of mountains as particularly sensitive indicators of global environmental change and as storehouses of natural resources was sufficiently established and continously growing (Messerli and Ives, 1997). In the same year, the IGBP published the report, *Predicting Global Change on Mountain Hydrology and Ecology* (Becker and Bugmann, 1997) and in 2001, the IGBP, IHDP and Global Terrestrial Observing System (GTOS) jointly published *Global Change and Mountain Regions* (Becker and Bugmann, 2001), which was the starting point for the Mountain Research Intitiative (MRI).

Looking back at the development of interdisciplinarity and transdiscplinarity, it is interesting to see 20 years later some remnants of the MAB model in the global change programmes. For example, Figure 2.9 shows the basic concept of the Land-Use and Land-Cover Change (LUCC) project (Turner et al, 1995), a joint project of IGBP and IHDP. As in the classical MAB model, the land-use system is in the centre, influenced and changed by the physical and the human driving forces. Impacts on the land-use system can be connected with biodiversity, water resources, food production, climate variability, health problems, urbanization and migration processes, exactly as in the MAB model. More important, however, is the fact that in all these recent publications the different definitions for interdisciplinarity and transdisciplinarity no longer exist. 'Interdisciplinarity' is often used to mean not only cooperation between different disciplines, but also the participation of actors outside acad-

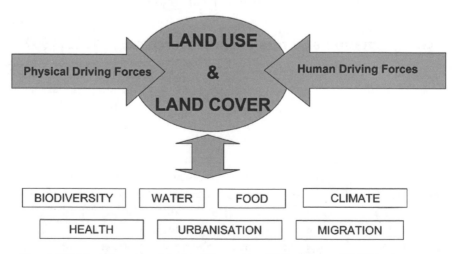

Figure 2.9 Adapted version of the Land-Use and Land-Cover Change (LUCC) basic concept

Source: Turner et al (1995)

emia in general, or with local population and authorities in particular. This simplification may show the overwhelming challenge of the global change programmes to overcome the gap between natural and human sciences, because disentangling the natural and human driving forces is one of the most demanding scientific problems, and solutions are not possible without cooperation, a common language and integrating approaches. This statement is especially important for mountain research and development, because natural and human processes are closely connected due to the fragility of mountain ecosystems.

The natural and social sciences have different cultures, and successful collaboration requires that each is aware of these differences. Without an open and flexible attitude and a willingness to learn on both sides, it will be easier to fail than to succeed. To achieve a certain level of integration, which is not the same as summation, requires an open-minded approach by both natural and social scientists – and especially also by the funding agencies (Gash, 2002). Therefore, for mountain projects, as for others, it is very important to keep in mind which factors are crucial for successful collaboration and which factors or behaviour will most probably lead to a failure. Box 2.2 gives an overview, even if this list of factors must be changed and adapted to every specific place and task.

Box 2.2 Some selected conditions for failed and successful interdisciplinary projects

Failed collaborations are likely to have:

- No shared concepts, no common goal
- Questions formulated by one side
- Problems with semantics
- Lack of commitment
- Misconceptions of roles and place
- Difficulty in attracting scientists
- Poor communication and physical separation
- Uncertain funding

Successful collaborations are likely to have:

- Shared concepts and language
- Excellence in own field
- Joint proposal development
- Sub-projects to allow individuals to succeed in their own field
- Intellectual respectability
- Long-term commitment
- Good communication and personal contact
- Safe funding

Source: adapted from Gash (2002)

One of the main issues in global change research is to look for the most appropriate methods to integrate knowledge from not only the natural and social sciences, but also the humanities, because most decisions about the environment include human values and beliefs. Understanding or predicting sustainability requires identification of the characteristics of resilient (durable) systems, and the dynamic relationship between knowledge production, policy formulation and decision-making (Wasson and Underval, 2002). Consequently, the global change programmes are using not only interdisciplinary, but also transdisciplinary approaches.

Conclusions

According to Steffen et al (2004):

> *The interactions between environmental change and human societies have a long and complex history, spanning many millennia. They vary greatly through time and from place to place. Despite these spatial and temporal differences, in recent years a global perspective has begun to emerge that forms the framework for a growing body of research within the environmental sciences. Crucial to the emergence of this perspective has been the dawning awareness of two fundamental aspects of the nature of the planet. The first is that the Earth itself is a single system, within which the biosphere is an active, essential component. Second, human activities are now so pervasive and profound in their consequences that they affect the Earth at a global scale in complex, interactive and accelerating ways; humans now have the capacity to alter the Earth system in ways that threaten the very processes and components, both biotic and abiotic, upon which humans depend.*

Therefore, systems approaches are necessary, from the local to the global scale; and these have rapidly developed in the last decades of the 20th century from the UNESCO MAB system through different interdisciplinary and transdisciplinary approaches to current global change research. Yet we should not forget that three main types of knowledge are necessary in any research for sustainability (see Figure 2.10) (CASS, ProClim, 1997):

* *system knowledge*: the current state of knowledge with its structures and processes, variabilities and interactions;
* *target knowledge*: knowledge concerning what may be or what should be. This requires an evaluation of the current situations to formulate prognoses and scenarios, providing critical levels or thresholds, guiding ideas, ethical boundaries, conditions, visions and so on. Science must stimulate discussions on the value and targets of future development;

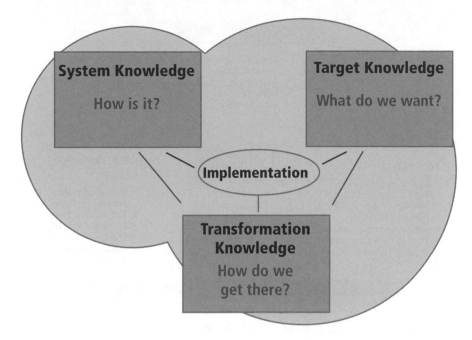

Figure 2.10 Research for sustainable knowledge needs three types of knowledge

Source: Cass, ProClim (1997)

- *transformation knowledge*: knowledge on how to shape and implement the transition from the existing to the target situation.

Future environmental and sustainability research needs to place greater emphasis on target and tranformation knowledge, especially in mountain research and development.

Reviewing the 25 years from local- to global-scale research, one topic remained the same in all the interdisciplinary and trandisciplinary approaches: the problem of collaboration between representatives of the natural and social sciences. This problem was recognized in the 'Declaration on Science and the Use of Scientific Knowledge' of the World Conference on Science (1999): 'Greater interdisciplinary efforts, involving both natural and social sciences, are a prerequisite for dealing with ethical, social, cultural, environmental, gender, economic and health issues'. In the same year, the *World Social Science Report* (UNESCO, 1999) stated: 'The convergence between natural and social sciences becomes greater to the degree one views both as dealing with complex systems, in which future developments are the outcome of temporally irreversible processes'.

'Global change' and 'globalization' are the catchwords of our time, but it is too often forgotten that the implications are very different for the developing and the developed world. Culture-specific land-use systems are being

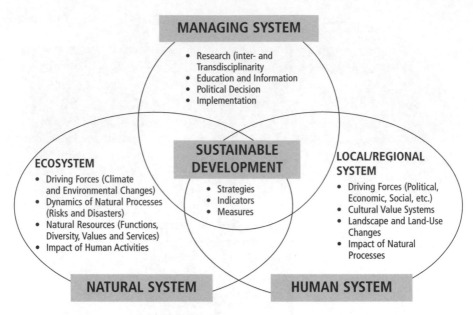

Figure 2.11 Key elements of disciplinary, interdisciplinary and transdisciplinary contributions to mountain research from the local to the global levels

wiped out because the priorities for research and development lie elsewhere in the industrialized sections of society. Lack of culture-specific technologies will lead to ecological catastrophes and large-scale social disruptions, as already evident in many traditional mountain societies in India (Ramakrishnan, 2001). This great divide is taken into consideration in Figure 2.11, which tries to show the development from the MAB model to the global environmental change approaches by integrating the natural and human systems with their specific resources and driving forces in connection with better management towards sustainability.

References

Becker, A. and Bugmann, H. (eds) (1997) 'Predicting global change impacts on mountain hydrology and ecology', *IGBP Report 43*, Stockholm, IGBP Secretariat, Royal Swedish Academy of Sciences

Becker, A. and Bugmann, H. (eds) (2001) 'Global change and mountain regions. The ountain Research Initiative', *IGBP Report 49, GTOS Report 28, IHDP Report 13*, Stockholm, IGBP Secretariat, Royal Swedish Academy of Sciences

Brugger, E. A., Furrer, G., Messerli, B. and Messerli, P. (eds) (1984) *The Transformation of Swiss Mountain Regions*, Berne, Haupt

Buttimer, A. (2004) 'Poetics, aesthetics and Humboldtean science', in Gamerith, W., Messerli, P., Meusburger, P. and Wanner, H. (eds) *Alpenwelt – Gebirgswelten, 54. Deutscher Geographentag*, Berne, Geographisches Institut, University of Bern

CASS, ProClim (1997) *Research on Sustainability and Global Change – Visions in Science Policy by Swiss Researchers*, Berne, Swiss Academy of Science

Centre for Our Common Future (1983) *The Earth Summit's Agenda for Change. A Plain Language Version of Agenda 21 and the Other Rio Agreements*, Geneva, Centre for Our Common Future

De Terra, H. (1955) *The Life and Times of Alexander von Humboldt 1769–1859*, New York, Alfred A. Knopf

Funnell, D. C. and Price, M. F. (2003) 'Mountain geography: A review', *Geographical Journal*, vol 3, pp83–190

Gash, J. (2002) 'Natural sciences, social sciences: Integration or summation?' *Global Change Newsletter*, vol 49, pp24–26

Humboldt, A. von (1805) *Essai sur la géographie des plantes, accompagné d'un tableau physique des régions équinoxiales, fondé sur les mesures exécutées depuis le dixième degré de latitude boréale jusqu'au dixième degré de latitude austral, pendant les années 1799, 1800, 1801, 1802 et 1803*, Paris

Humboldt, A. von. (1845–1862) *Kosmos. Entwurf einer physischen Weltbeschreibung, 5 Bde und ein Atlas, hrsg. von Traugott Bromme 1851 und Carl Troll 1961*, Stuttgart

Hurni, H., Wiesmann, U. and Schertenleib, R. (2004) *Research for Mitigating Syndromes of Global Change. A Transdisciplinary Appraisal of Selected Regions of the World to Prepare Development-Oriented Research Partnerships. Perspectives of the NCCR North-South*, Berne, Geographica Bernensia

Ives, J. D. and Messerli, B. (1990) 'Progress in theoretical and applied mountain research, 1973–1989, and major future needs', *Mountain Research and Development*, vol 10, pp101–127

Messerli, B. (1980) 'Die afrikanischen Hochgebirge und die Klimageschichte Afrikas in den letzten 20,000 Jahren', in Oeschger, H., Messerli, B. and Svilar, M. (eds) *Das Klima*, Berlin, Springer

Messerli, B. and Hurni, H. (eds) (1990) *African Mountains and Highlands: Problems and Perspectives*, Berne, Geographical Institute, University of Berne

Messerli, B. and Ives, J. D. (eds) (1997) *Mountains of the World: A Global Priority*, Carnforth and New York, Parthenon

Messerli, B. and Messerli, P. (1978) 'Wirtschaftliche Entwicklung und ökologische Belastbarkeit im Berggebiet', *Geographica Helvetica*, vol 4, pp203–210

Messerli, B., Hurni, H., Wolde-Semayat, B., Tedla, S., Ives, J. D. and Wolde-Mariam, M. (1988) 'African mountains and highlands', *Mountain Research and Development*, vol 2/3, pp93–100

Messerli, P. (1983) 'The concept of stability and instability of mountain ecosystems derived from the Swiss MAB-6 studies of the Aletsch area', *Mountain Research and Development*, vol 3, pp281–290

Messerli, P. (1986) *Modelle und Methoden zur Analyse der Mensch-Umwelt-Beziehungen im alpinen Lebens- und Erholungsraum*, Erkenntnisse und Folgerungen aus dem schweizerischen MAB-Programm 1979–1985, Berne, Bundesamt für Umweltschutz

Messerli, P. (1989) *Mensch und Natur im alpinen Lebensraum, Risiken, Chancen, Perspektiven*, Berne, Haupt

Messerli, P., Liniger, H. P. and Müller, I. (2003) *Tourismus, Berglandwirtschaft und Naturschutz – Die Entwicklung einer nachhaltigen Beziehung im Aletschgebiet*, Exkursionsführer, Berne, Geographisches Institut, University of Bern

Messerli, P. D., Wiesmann, U. and Hurni, H. (2005) 'The mountains and highlands focus of the Swiss National Centre of Competence in Research (NCCR) North-South', *Mountain Research and Development*, vol 25, pp174–179

Moser, W. (1975) *Einige Erfahrungen mit dem Tourismus in den Alpen – das Oekosystem Obergurgl*, Munich, Schriftenreihe des Alpeninstitutes München

Patzelt, G. (ed.) (1987) *MAB-Projekt Obergurgl*, Veröffentlichungen des Oesterreichischen MAB-Programms 10, Innsbruck, Universitätsverlag Wagner

Price, M. F. (1995) *Mountain Research in Europe. An Overview of MAB Research from the Pyrenees to Siberia*, Carnforth, Parthenon

Ramakrishnan, P. S. (2001) *Ecology and Sustainable Development*, New Delhi, National Book Trust

Steffen, W., Sanderson, A., Tyson, P. D., Jaeger, J., Matson, P. A., Moore, B., Oldfield, F., Richardson, K., Schellnhuber, H. J., Turner, B. L. and Wasson, R. J. (2004) *Global Change and the Earth System: A Planet under Pressure*, Stockholm, IGBP Secretariat

Thompson Klein, J., Grossenbacher, W., Häberli, R., Bill, A., Scholz, R. W. and Welti, M. (eds) (2001) *Transdisciplinarity: Joint Problem Solving Among Science, Technology and Society. An Effective Way for Managing Complexity*, Basel, Boston andBerlin, Birkhäuser

Turner, B. L. II, Skole, D., Sanderson, S., Fischer, G., Fresco, L. and Leemans, R. (1995) 'Land-use and land-cover change. Science/Research Plan', *IGBP Report 35/IHDP Report 7*, Stockholm, IGBP Secretariat

UNESCO (1999) *World Social Science Report*, Paris, UNESCO Publishing/Elsevier

Wasson, B. and Underdal, A. (2002) 'Human-environment interactions: Methods and theory', *Global Change Newsletter*, vol 49, pp22–23

Wiesmann, U. (2001) 'Umwelt, Landwirtschaft und Tourismus im Berggebiet – Konfliktbearbeitung im Leitbild "Grindelwald 2000"', in Oesterreichisches Studienzentrum für Friedens- und Konfliktforschung (eds) *Konfliktbearbeitung und Kooperation*, vol 7, Münster, Agenda Verlag

World Conference on Science (1999) *Science for the Twenty-First Century: A New Commitment*, Paris, UNESCO, ICSU

3

From Subsistence Cultures to Sustainable Development: Linking Knowledge Systems for *Jhum*-Centred Land-Use Management in Northeast India

P. S. Ramakrishnan

Introduction

Forest farmers in India's northeastern hill region, as elsewhere in the tropics (Nye and Greenland, 1960; Ruthenberg, 1971; Spencer, 1966; UNESCO, 1983), have sustainably managed their traditional shifting agriculture over the centuries using 'slash-and-burn agriculture', locally known in India as *jhum*. This is essentially an agroforestry system organized in both space and time. In the past, the small-scale perturbations under longer *jhum* cycles (the length of the forest fallow phase between two successive cycles on the same site) ensured enhanced biodiversity in the forest, with human-enriched crop and associated biodiversity; long cycles also ensured a rich nutrient capital in the soil released through slash and burn. With increasing pressure on forest resources from both outside and within the region, the *jhum* cycle has come down to a short five years or less (except for longer cycles still found in remote areas), with consequent land degradation and decline in soil fertility. The intensification of agriculture in many situations has led to rotational-fallow systems with little or no burning of biomass, ending up in sedentary agricultural practices.

In addition to *jhum* as the major land use, there exists in the region a whole variety of other agroecosystems: home-garden systems, traditional cash-crop plantation systems, and extensively distributed valley wet rice cultivation systems at varied levels of intensification. It was also realized early on during the initiative described in this chapter, which started in the early 1970s, that many of the major agroecosystem typologies have sub-typologies that are society-specific and adapted to diverse ecological conditions. The forest ecosystem, at varied levels of degradation, has also been a source of societal benefits, both tangible (Gangwar and Ramakrishnan, 1990) and intangible

(Khiewtam and Ramakrishnan, 1989). Therefore, we need to take a functional perspective focusing on the village ecosystem, with all its subsystems – agriculture, animal husbandry and domestic subsystems – as the focal point (Maikhuri and Ramakrishnan, 1990; Mishra and Ramakrishnan, 1982). We must also recognize that the landscape is sculptured in a variety of different ways, depending upon the eco-cultural attributes of the given ethnic group(s) living and/or coexisting in the area, sharing natural resources around them in a sustainable manner. Thus each ethnic society has given the 'natural cultural landscape' a distinct identity, based upon their perceptions and traditional value systems (Ramakrishnan et al, 1998, 2005).

The project

Land degradation associated with shortening of the *jhum* cycle has various stages: arrested succession of bamboos under 10–30 year cycles; takeover by weeds under cycles of less than 10 years, often dominated by invasive species; and further degradation ending up in a balded and desertified landscape – in a region receiving an average rainfall of 200cm during the monsoon, and an average annual rainfall that may reach 12m and, in an exceptional year, 24m (Ramakrishnan, 1992a). Many studies on areas affected by shifting agriculture tend to imply that shifting agricultural farmers are responsible for land degradation. In the late 1980s, it was estimated that about 500 million people were dependent upon shifting agriculture on c. 400 million hectares of tropical forest land in 90 countries (FAO/UNEP, 1982). More than half of the 7 per cent change resulting in moderate to severe land degradation, observed from 1980 to 1990, was suggested to be due to shifting agriculture (FAO, 1995), and this was often cited as a reason to get rid of *jhum* through alternative land uses.

Not convinced with the line of thinking that *jhum* farmers are responsible for land degradation – which implied that they are, in a sense, inflicting damage on their own land and the biodiversity contained therein, with implications for their own livelihood needs – we looked at the dynamics of land-use and land-cover change. From this also arose our efforts to look at location- and ethnicity-specific traditional ecological knowledge (TEK) carried forward through an experiential process spread across generations. The objective was to document this knowledge base and the way farmers determine ecological processes, thus helping them make land-use management decisions. This implied that we should not only understand the location-specificity of this knowledge, but also be able to make generalizations across socio-ecological systems so that research analysis fitted into a regional developmental planning process. The ultimate objective was to effectively build developmental technologies upon TEK, with appropriate inputs from the 'formal' knowledge system related to ecosystem structure and function.

Dubbing *jhum* as a 'primitive' land-use practice, over the last 100 years governmental agencies have tried to replace it with alternative technologies such as sedentary terraced farming, even providing incentives for such change. The farmers have time and again rejected this. Meanwhile, they were forced to operate under rapidly shortening *jhum* cycles that are neither ecologically nor economically efficient, in spite of some adaptive changes they made in their cropping systems. The net result is to reduce the system's stability and resilience, which are essential for coping with the ecological uncertainties common in a mountain environment. Therefore, one of the concerns at the time when this interdisciplinary study began was to examine land-use sustainability issues by designing a community participatory developmental pathway based on a value system that the farmers understand and appreciate. A corollary to this was also to have a community-based biodiversity conservation strategy, both natural and human-managed, in this 'hotspot' region of the world, a philosophy that started emerging through the report of the World Commission on Environment and Development (1987).

Having successfully gone through the first phase of this project initiative over more than 20 years from the early 1970s through to the 1990s, we have extended our studies to other mountain regions of India – the Himalaya and Western Ghats – with an initial learning-through-working phase of a couple of years at the beginning. Coming from a traditional biophysical ecology background, the challenges of getting into interdisciplinary studies involving well over 100 socio-ecological systems, each one unique in its own right, was a stupendous task – let alone arriving at generalized regional conclusions for the design of land-use development. Looking back, it is now indeed satisfying to see that an intensely participatory interdisciplinary research initiative, with limited funding (US$10,000–20,000) coming from a number of different Indian sources, could have a multiplier effect, eventually having a major impact at the regional level, and creating awareness of the problems of traditional mountain societies at the national level. Indeed, it triggered a developmental paradigm for the traditional societies that is generated from within, rather than imposed from outside; a paradigm based on a value system that local communities understand and appreciate, so that they could participate in the whole process of a research-based developmental initiative.

Interdisciplinarity

While the initiative began as a multidisciplinary analysis involving ecological and social science disciplines, this collaborative relationship was soon abandoned because of the obvious 'cut and paste' possibilities arising from different methodologies. Abandoning the questionnaire approach taken by the social scientists, we soon realized that only an intensive participatory approach, working year-round with local communities, would help to analyse and validate TEK, since traditional societies themselves cannot articulate the knowledge that links ecological and social processes. Very soon we realized that TEK forms a powerful connecting link, and that this knowledge system has to be appropriately linked with textbook-based 'formal' knowledge to create synergies between ecological conservation and the development of traditional societies. One other objective in this initiative was to create a new breed of ecologists with the ability to work in the interface between natural and social sciences. Involving over 30 PhD-level workers and an equal number of post-doctoral researchers, and with eight to ten community participants for each of them, all the diverse elements of the unique cultural landscape(s) created by diverse ethnic groups were investigated through a socio-ecological system approach (ecological, socio-economic and cultural), addressing the linkages that exist between natural and human-managed ecosystems and looking at issues at scales varying from the sub-specific and species levels to ecosystem (sacred grove) and landscape (cultural landscape) levels.

Integrating knowledge systems for sustainability

The formal knowledge system in ecological sciences is based on an analysis of the biophysical principles underlying the organization and functions of an ecosystem. Derived through a hypothetico-deductive process, perturbation is now seen as the driving force determining ecosystem organization and functioning, with implications for our understanding of successional patterns and processes such as population dynamics, biomass and production functions, energy flow, hydrology and soil fertility dynamics, and nutrient cycling. Such an approach leads to an understanding of the variety of typologies of ecosystems (both natural and human-managed), the organization of species in space and time and their functional attributes, and factors contributing to arrested succession and/or biological invasion, all eventually leading to land degradation and site desertification, with implications for the ecosystem-level process (Ramakrishnan and Vitousek, 1989) – but delinked from the human dimensions.

When dealing with traditional societies largely confined to the mountains of the developing tropics, integration in the real sense is best achieved through a knowledge system to which these communities are able to relate. This is where the community-based TEK, derived from an experiential process of societal interaction with nature and natural resources, becomes important for addressing ecosystem sustainability concerns through community participation. This implies that TEK has a certain degree of location-specificity, in an area with over 100 different ethnic societies each with their own language, culture and customs, and with a strong human element attached to these that emphasizes social emancipation (Elzinga, 1996). The challenge from a policy perspective has always been to arrive at generalizations that are applicable across socio-ecological systems, so that regional developmental concerns could be addressed, and at the same time taking on board socio-ecological diversities existing in the region.

On the basis of the limited validated information that is now available, the benefits accruing from TEK may be viewed as three kinds (Ramakrishnan, 1996; Ramakrishnan et al, 1998):

1 economic – traditional crop varieties and lesser known plants and animals of food, medicinal or other value harvested from the wild;
2 ecological/social – manipulation of above- or below-ground biodiversity for coping with uncertainties in the environment and global change, controlling soil water regimes and hydrology, efficient organic residue management, and soil fertility management through modified soil biological processes;
3 ethical – cultural, spiritual and religious belief systems centred around the concept of the sacred species, sacred groves and sacred landscapes.

We had to move beyond all these perceptional divergences between knowledge systems to get an integrated picture of the landscape organization and functions. By appropriately linking TEK with 'formal' knowledge, not only have we been able to translate research results into policy formulations for sustainable natural-resource management linked with the development of shifting agricultural communities in northeastern India, we have also been able to follow this pathway in many other mountain societies of the developing tropics (Ramakrishnan, 2001; Ramakrishnan et al, 2005).

Stakeholder involvement

Living and working with village communities year-round, over an extended period of up to two years, ensured the intensive participation of the most important stakeholders, namely the local communities, with the active involvement of the village headmen who play key roles in influencing local decision-making processes. From the researcher's viewpoint, the important objective was to understand the science behind *jhum*; from the villager's viewpoint, the objective was an improved quality of life made possible through appropriate institutional arrangements, taking on board both traditional and modern ways of institution building for effective community participation. Constant interaction with the textbook knowledge of the agricultural scientific community, the policy planners and developmental agencies led to the realization that implementing the emerging approach towards building upon *jhum*, rather than finding an 'alternative to *jhum*', is indeed an uphill task.

Research results

Realizing that the *jhum* farmers are held responsible for forest conversions and the land degradation that is linked to it, we investigated the causative factors for the rapidly shortening shifting agricultural cycles through socio-ecological analysis of the problem of land degradation. Our conclusion was that over the last 100 years, external pressures from timber extractors from outside are responsible for large-scale deforestation and the consequent rapid land degradation. With limited choices available to restrict *jhum* activity to less degraded sites, and increasing local population pressure acting as a proximate factor, the *jhum* farmer had to operate under reduced cycle lengths (Ramakrishnan, 1992a). This, along with governmental policies and market forces, exacerbated land degradation. This interplay between proximate and key drivers leading to land degradation has now started to receive greater attention (Lambin et al, 2001), with implications for addressing sustainability concerns. Indeed, as discussed below, sacred groves themselves are indicative of the value that traditional societies attach to good forest cover, though in the name of so-called 'modernization', such value systems have started to fall apart!

What follows is an integrated view of the traditional and the formal knowledge systems, as part of a cultural landscape unit in which natural and human-managed systems coexist, with the village ecosystem forming the connecting link for the cultural landscape.

Forest ecosystem

Traditional knowledge: Intangible benefits with tangible outcomes

Traditionally being animistic in their religious beliefs, *jhum* farmers view nature and the natural resources around them with respect and reverence. As an expression of this value system, they tend to have a set of rules and rituals, interpreted through the village elder or the priest, that may lead to positive impacts or taboos that can cause disaster. Indeed, the *jhum* farmers, irrespective of their ethnic affiliation and location, often have a *jhum* calendar right from the stage of slashing and burning of the forest, through seeding and final harvest(s) of the crop yield from their mixed cropping systems, which is linked with socio-cultural festivals and ceremonies – the expressions may vary but the psychological basis remains. To cite one such example, the Garos of Meghalaya believe that the first Garo to settle on their land was 'Bone Nirepa Jane Nitepa', who started *jhum* with blessings from the deity, Misipa. Soon after land allocation, at the time of slashing, Garos perform a religious ritual, followed by a series of rites: *algalmaka* at the time of burning and planting of the crops; *miamua* performed by Nokma (the village representative of the land-owning clan) when the rice is fruiting; *rongchuygala* and *ahia*, linked with some of the vegetable crops that mature. The culmination of the *jhum* activity itself is marked by the famous drum festival, *wangala*, performed in honour of Saijong, the sun god, who is the ultimate protector of the crops.

While these intangible benefits are directly linked with *jhum* activity itself, as an expression of their respect for nature, *jhum* farmers have protected specific ecosystem types (sacred groves) for cultural/religious reasons. This tradition of having a sacred grove for each village, with many religious ceremonies performed within the grove during the year in order to propitiate natural elements and the ruling deity, is directed to guard against famine and epidemic diseases (Khiewtam and Ramakrishnan, 1989), though these traditions are being eroded under modern influences. The details regarding the institutional arrangements for managing the groves through rituals and restrictions imposed on the society by the priest vary from place to place. Often, these groves remain as islands of biodiversity in a largely degraded landscape. Indeed, the concept of the sacred grove can be seen as a precursor of the concept of nature reserves that are currently managed for the protection of a given endangered species or set of species (Ramakrishnan, 1992a).

The Mawsmai grove in Cherrapunji and the Mawphlong grove near Shillong are two examples of rigorously protected groves in Meghalaya. In a more proactive view of conservation, the restoration of adjacent degraded ecological systems becomes an important dimension, a developmental aspect that is strengthened through an understanding of the structural (spatial distribution of species and their biomass) and functional (productivity, soil fertility, hydrological processes and nutrient cycling patterns) attributes of this protected system (Khiewtam and Ramakrishnan, 1993). The biophysical ecological studies in the Mawsmai sacred grove provide rich insights into the tangible benefits accruing to society from this very fragile sacred grove. Interestingly, in this high rainfall spot of the world, where the average annual precipitation is over 12m, rising to 24m in an exceptional year, as in 1974 at Cherrapunji, the dense rainforest is confined to the sacred grove only, with xerophytic grass formations elsewhere, often ending up in a balded desertified landscape. The causative factor for this is the short-cycling of nutrients through a surface root mat sitting on an almost non-existent soil profile (Khiewtam and Ramakrishnan, 1993).

Many species that are socially valued within and outside the sacred grove often play a keystone value within the ecosystem. Many early successional species such as Nepalese alder (*Alnus nepalensis*) and many bamboo species come under this category. Nepalese alder can conserve as much as 125kg/ha each year[1] of nitrogen within the system where it is present (Ramakrishnan, 1992a). Similarly, many socially valued bamboo species can conserve key elements such as nitrogen, phosphorus or potassium, depending upon the species. Species such as the broad-leaved *Englehardtia spicata*, *Echinocarpus dasycarpus*, *Sysygium cuminii* and *Drimycarpus racemosus*, found on the highly infertile soils of the Mawsmai sacred grove, conserve high levels of nitrogen, phosphorus and potassium, as much as around a third of what is conserved in this dense rainforest ecosystem (Ramakrishnan, 1992a). More extensive stud-ies have shown that socially valued species are also ecological keystone species in other parts of the world (Ramakrishnan et al, 1998). Thus, one or more keystone species within the system can not only determine ecosystem processes but also impact upon the structural and functional attributes of the system at the landscape level. Selecting such species for forest fallow management linked to *jhum* would ensure community participation, since local people are able to relate to a value system that they understand.

Formal knowledge for forest management

Working with the *jhum*-centred forest ecosystems at various levels of degrada-tion, we have undertaken integrated analyses of forest successional patterns and processes linked to soil fertility dynamics and nutrient cycling processes (Ramakrishnan, 1989a). These address issues such as adaptive strategies of species and populations, including factors contributing to arrested succession of bamboos (Ramakrishnan, 1989b; Rao and Ramakrishnan, 1989) or the spread of native and exotic weeds (Ramakrishnan, 1986, 1989a, 1989b, 1991). We have also looked at the soil subsystem over a successional gradient, through an understanding of earthworm population dynamics (Bhadauria and Ramakrishnan, 1989, 1991, 1996); people see earthworm species as indicators of soil fertility, and so they are also socially valued. Through such detailed analyses of *jhum*-centred forest ecosystem structure and functions organized in space and time, we have been able to arrive at conclusions on soil fertility and nutrient cycling linked to fallow management of the already weakened *jhum* under short cycles of less than ten years (Ramakrishnan, 1992a, 2001). In any case, natural or assisted regeneration of the forest fallow reduces nutrient loss and allows a return to the steady-state cycling characteristics of mature forest (Toky and Ramakrishnan, 1983a, 1983b).

At another level of understanding successional patterns and processes, research on the growth pattern and architectural analysis of the shoot and root systems of early versus late successional tree species enabled us to identify tree species that have rapid growth and also the right architectural features to be used in fallow management for a redeveloped *jhum*. This study looked at shoot architectural design through an analysis of the extension growth of the main axis, branching patterns and growth of different orders of branches, root architectural design at different soil depths, and lead dynamics through an analysis of the birth, death and turnover rates of leaves, considering leaves as a meta-population, over a successional gradient (Boojh and Ramakrishnan, 1982a, 1982b; Shukla and Ramakrishnan, 1984a, 1984b, 1986). The aim was to look at complementarities of late versus early successional tree architecture for effective light capture, and of the root architecture for nutrient capture from the soil profile, through a condensed succession in a mixed plantation programme (Ramakrishnan, 1986).

We need to integrate the traditional forester's way of looking at forestry as a silvicultural activity with a whole variety of other dimensions of the problem (see Figure 3.1). Ecological knowledge from areas such as tree biology and architecture, patch dynamics, ecophysiology of developing forest communities, reproductive biology and nutrient cycling processes could all be integrated into the current management process and future management options. To ensure community participation, we need to bring in the social and cultural dimensions through an in-depth understanding of the tangible and intangible values of the forest ecosystem through the TEK base available in the given situation, appropriately linking this with the formal knowledge-based under-standing of the forest ecosystem function.

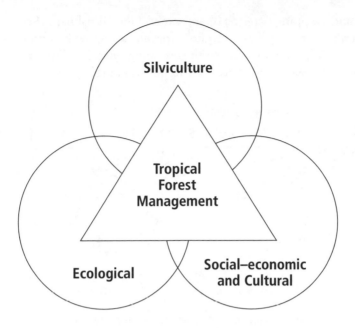

Figure 3.1 Interdisciplinary interactions called for in tropical forest management and conservation

Source: Ramakrishnan (1992b)

The *jhum* system

Formal knowledge centred on jhum

After the forest is clear-cut and burned, the ecosystem loses its ability to hold nutrients (Ramakrishnan, 1989a). Losses occur through the volatilization of carbon (C) and nitrogen (N) during the burn. Substantial nutrient losses through wind-blow of ash, runoff and leaching through water may all occur before adequate vegetal cover builds up, during both the cropping and fallow phases. During the cropping phase, losses also occur through the uptake and removal of nutrients in biomass that is harvested. Hydrological measurements and the closely linked soil fertility budget measurements suggest that, with the shortening of the *jhum* cycle to below ten years, the system tends to become unstable (Toky and Ramakrishnan, 1981, 1983a, 1983b). Rapid regeneration of forest vegetation following clearing and burning reduces nutrient loss and allows a return to the steady-state cycling characteristics of mature forest.

With the cycle length coming down to an average of five years or even less, crop yield declines sharply because of poorer recovery of soil fertility (Ramakrishnan and Toky, 1981) and partly due to increased weed potential (Saxena and Ramakrishnan, 1984). Although farmers deal with a decline in soil fertility under cycle lengths of less than five years through alterations in crop mixtures and cropping patterns (Gangwar and Ramakrishnan, 1987), normally a minimum of 10 to 15 years is required for fallow regrowth in order to recoup most of the soil fertility lost during the cropping phase, as illustrated in Figure 3.2 for the elements phosphorus (P) and potassium (K).

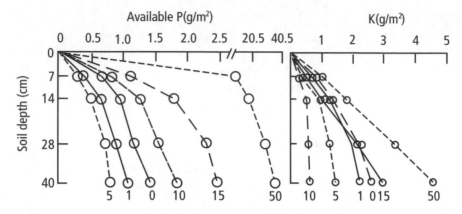

Figure 3.2 Changes in cumulative quantity of available phosphorus and potassium within a soil column of 40cm depth after *jhum* fallow periods of 0, 1, 5, 10, 15 and 50 years

Source: Ramakrishnan and Toky (1981)

However, through the presence of socially selected species such as Nepalese alder (*Alnus nepalensis*) in the *jhum* plots, the farmer is able to conserve a key nutrient such as nitrogen, making even a shorter *jhum* cycle of five years sustainable. Nitrogen budgets under different cycle lengths of 15, 10 and 5 years are illustrative of the kind of issues involved (Mishra and Ramakrishnan, 1984). During one cropping phase, the agroecosystem loses about 600kg/ha of N (the difference between the soil N capital before and after one cropping). With the plot under a five-year cycle, having the same cycle length during the last 20 years, this system had lost 1.28 × 10³kg/ha of N from its initial capital of 7.68–6.40 × 10³kg/ha. While 10- and 15-year agricultural cycles are long enough to restore the original N status in the soil before the next cropping, it seems unlikely that, except for the presence of a keystone species like Nepalese alder, the 600kg/ha of N lost during one cropping could be restored under a five-year cycle. Similar studies are also available for potassium budgets (Swamy and Ramakrishnan, 1988).

One reason why *jhum* is able to persist, irrespective of all governmental initiatives to the contrary, is because it is a low-input system based on organic residue management, with labour as the main energy input and a high energy efficiency, with an output/input ratio between 15 and 50 (Ramakrishnan, 1992a). When the *jhum* cycle comes down from the long 60-year cycle – still encountered in more remote areas – to an average cycle of around five years, both soil fertility status (Ramakrishanan and Toky, 1981) and the consequent reduced economic yield decline (see Figure 3.3). A major conclusion that emerged from agroecosystem analysis was that a minimum of a 10-year cycle was optimal from the sustainability perspective. This was reflected in the economic efficiency measurements through monetary output/input analysis. However, from the point of view of forest management, a long fallow period of 30 years or more is required to build a good forest cover (Toky and Ramakrishnan, 1983b).

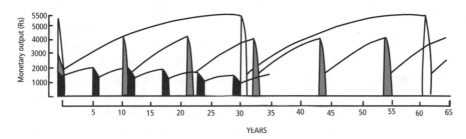

Figure 3.3 Economic yield pattern under different shifting agricultural cycles of 30 (white bars), 10 (grey bars) and 5 (black bars) years in northeast India

Source: Ramakrishnan (1992a)

Less environmentally friendly is a situation where further shortening of the cycle, leading to an arrested succession of weedy species, is complicated further by large-scale invasion of exotic weeds such as *Eupatorium* and *Mikenia* from an alien environment, in this case South America (Ramakrishnan, 1991). Extensive weed takeover is difficult for the *jhum* farmer to deal with. In this context, tree growth strategies based on formal knowledge, and particularly the socially valued ecologically significant keystone species based on TEK discussed earlier, are tools of value for fallow management of forest and a redeveloped *jhum* system.

Traditional knowledge centred on jhum

Built around a rich traditional knowledge base, *jhum* is a highly complex multi-cropping system organized in both space and time, with implications for economic yield from a given site (Ramakrishnan, 1992a). Emphasizing fewer (5–10) species that are nutrient-use-efficient, such as tuber and vegetable crops, under shorter *jhum* cycles, and having a much larger number of species (up to 40–45), including less-efficient ones, under longer cycles, the farmer can make use of the limited available nutrients that are transient on a hill slope of about 40°. While all the crop species are sown together soon after the first few monsoon rains, crops are harvested sequentially over a period of a few months, so that continuous crop cover is maintained. Indeed, such sequential harvesting not only provides additional space for the remaining crops, but also reduces competitive pressures. Continually recycling the harvested biomass also improves soil quality throughout the growing season. Indeed, the productivity of the *jhum* system is also linked to the level of sophistication in TEK of different ethnic groups (Maikhuri and Ramakrishnan, 1990, 1991).

As the *jhum* cycle comes down below five years, due to increased pressure on the land around city centres, with market forces determining land-use activities, there has been a gradual shift from the typical *jhum* to a rotational fallow management system with only a low burning of weedy biomass. This eventually leads to a sedentary system in response to increased population pressure, market forces and soil fertility decline. A lesser-known legume, and a keystone crop species such as *Flemingia vestita*, is brought into these cropping systems as an evolving adaptation in TEK to ensure sustainable soil fertility (see Figure 3.4). *F. vestita* can fix up to 250kg/ha of N each year (Gangwar and Ramakrishnan, 1989a), stabilizing *jhum* at very short cycle lengths and enabling traditional societies, through TEK, to cope with uncertainties in the environment under such conditions.

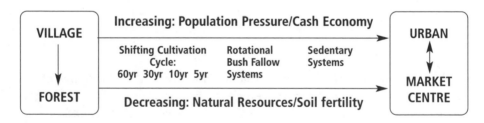

Figure 3.4 Land use changes as related to population pressure, land degradation and available linkages to a market economy

Note: These pressures decline with distance from the urban centre in northeast Indian uplands. Also note that the *jhum* system is replaced by a rotational fallow system leading to sedentary farming with decreasing distance from the urban centre

Source: Adapted from Ramakrishnan (1992a)

As a functional group, weeds form an integral part of the traditional *jhum* agroecosystem function. The 'non-weed' concept, through which the *jhum* farmer practices weed management rather than weed control (Swamy and Ramakrishnan, 1987), thus leaving around 20 per cent weed biomass *in situ* without being pulled out and recycling the pulled-out biomass back into the plot, is an interesting way of optimizing resource use. Indeed, this weed biomass not only does not compete with the crops, but contributes towards conserving nutrients that might otherwise be lost through hydrological processes of runoff and leaching. Indeed, 20 per cent weed biomass is the cut-off point when a weed becomes a non-weed in the cropping system, a practice also found in Mayan agriculture in Mexico (Gliessman, 1988). This is an example of TEK, evolved perhaps through many generations of experience, with significant soil fertility conservation implications.

Other land-use systems linked to *jhum*

Although *jhum* is the major land use in the region, a whole variety of land-use systems are part of its forested landscape. Only some of the more important are considered here.

Home gardens and traditional plantation systems

Of all the land-use systems linked to *jhum*, 'home gardens' play a significant role in meeting a variety of needs of traditional societies in the region and elsewhere in the developing tropics (Gliessman, 1990; Ramakrishnan, 1992a). Also known as kitchen gardens or forest gardens, they show considerable structural variations, functional differences and spatial complexity, depending upon the ecological and social settings in which they occur. Imitating a natural forest with a highly stratified and compacted set of economically important trees, shrubs and herbs in small plots of 0.5–2ha, these gardens may have over 100 species, providing the farmer with many of his requirements – food, fodder, firewood, spices, ornamentals and medicinal plants – all year round. These systems are ecologically and economically very efficient and of multipurpose value to local communities (Ramakrishnan, 1992a) and, once established, demand only very casual management (Swift et al, 1996).

Under declining soil fertility and site degradation, as in Cherrapunji, the shift from shifting agriculture to traditional plantation systems based on home gardens is an obvious phenomenon, which could be viewed as a human response to changing conditions, organized in both space and time (Gangwar and Ramakrishnan, 1989b) and based on evolving TEK. Such land-use changes could be seen as a response to changing biophysical conditions, increasing population, labour constraints, available off-farm employment opportunities, and insurance against possible loss of outside income. The home gardens and traditional plantation systems provide an additional window to find a meaningful solution to *jhum* by taking the pressure off the latter, through a diversified cash-crop economy.

Valley wet rice cultivation

A major land-use system available in the widespread valley lands is the sedentary wet rice valley cultivation system, which is closely linked to *jhum* on the hill slopes because the valleys receive nutrient wash-outs (Ramakrishnan,

1992a). Generally speaking, there are considerable variations in both the ecological efficiency measure of energy output/input ratio, which ranges from 2 to 18 for most communities, and the economic output per hectare, which does not exceed Rs3000–6000 (about US$70–140), in spite of heavy labour input – depending upon the socio-ecological conditions and the TEK-based wet rice cultivation system available to the communities (Ramakrishnan, 1992a). As there is often only one harvest during the rainy season, there is scope for improving the efficiencies of these systems through appropriate water and nutrient management strategies.

An interesting example of the possibilities for improving valley rice cultivation is the outstanding example of the highly evolved wet rice cultivation of the ethnic society of the Apatanis, confined to valley land at higher elevations of Arunachal Pradesh (Kumar and Ramakrishnan, 1990). With high variability within the system, the Apatanis have chosen traditional crop varieties of rice for a nutrient-rich soil, which differ from those chosen for a nutrient-poor soil. Often, four or five traditional varieties are grown together. With an elaborate water and nutrient management plan that they have evolved over time, they not only generate enough food for their own consumption, but also sell rice in the market. This highly organized wet rice cultivation has a high level of ecological efficiency, with over 70 units of energy output per unit input and, at the same time, has a comparable economic efficiency to that of the Green Revolution modern agriculture of the northwestern plains of India, touching Rs70,000 (US$1650) per hectare, with labour as the only input, as in all other wet rice cultivation systems in the region. Indeed, the Apatanis' wet rice cultivation system, with synchronized pisciculture embedded within it, offers immense opportunities for further strengthening this land-use system based on their TEK-based traditional technology, which decreases pressure on *jhum*.

The village ecosystem as part of a cultural landscape

As discussed above, traditional societies living in India's northeastern hill areas try to ensure self-sufficiency within their village ecosystem through a variety of traditional agricultural and animal husbandry activities, for which they are dependent upon a vast forest resource base around them (see Figure 3.5) as part of a cultural landscape (Ramakrishnan et al, 2005). Traditional food production systems are less energy-intensive, largely dependent upon resource recycling from within the surrounding landscape. In a sense, therefore, the complex agroecosystems discussed here are based on the background information that the surrounding landscape has to offer, with TEK being adapted in both space and time, depending upon changes in the biotic composition, natural or human-managed, in response to local necessities or modified goals (Ramakrishnan, 2001; Ramakrishnan et al, 2005). Thus, at one extreme are communities obtaining much of their daily needs as hunter-gatherers, for example, the Sulungs of Arunachal Pradesh, with other communities having

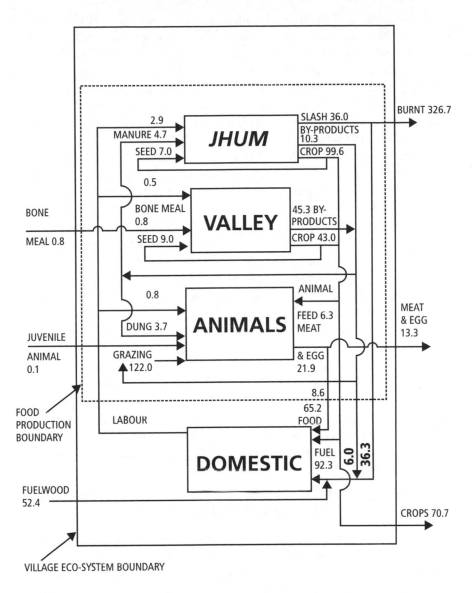

Figure 3.5 Interconnections between agriculture, animal husbandry
and domestic sectors, as indicated through energy flow (MJ × 10³)
in a Khasi village ecosystem in Meghalaya, northeast India

Source: Mishra and Ramakrishnan (1982)

increasing levels of sophistication in their agricultural and animal husbandry
practices with *jhum* as the main land use, leading to the highly evolved sedentary agricultural systems such as that of the Apatanis (Ramakrishnan, 1992a) –
though it is difficult to place them along a linear gradient.

Depending on labour as the main energy input, and with recycling of resources between agriculture, animal husbandry and domestic subsytems, linked to the forested landscape around (Ramakrishnan, 1992a), TEK determines the level of integration of the humans within the landscape boundaries. The way traditional societies have built up linkages between natural and human-managed ecosystems using TEK as a driver has, in the past, contributed to the overall sustainability of the landscape level, ensuring socio-ecological system integrity at the same time.

At its 16th session, the World Heritage Committee recognized the associative values of landscapes and landscape features to indigenous people, and the importance of protecting biological diversity through cultural diversity within cultural landscapes (Rossler, 2001). Such 'associative landscapes', that are continually evolving and operating at the cutting edge of ecology, economics and ethics, are becoming more and more relevant for natural resource management that is also concerned with the sustainable livelihood of local communities. In the contemporary context, these cultural entities provide not only intangible benefits that enable humans to arrive at a harmonious relationship with nature, which includes leisure, but also tangible benefits through the biodiversity that is conserved and managed through human actions. Our research initiatives aim to conserve or redevelop these cultural landscapes, providing an improved quality of life to the local communities. The objective of the sustainable management of natural resources, contributing to sustainable livelihoods and the development of traditional mountain societies of the developing tropics, is only possible through an understanding of the ecological and social processes at different scales (see Figure 3.6); TEK is an important connecting link and therefore plays a key role in this effort.

Taking research results into policy dimensions and developmental initiatives

Taking research results into the policy and developmental arena has not been an easy task. Using the spoken, written and audio-visual media, a developmental plan that essentially implied that 'if you cannot replace something with an alternative system, why not work with the local communities to build upon what is available', was the philosophy advocated for more than 20 years before it found acceptance with the governmental and funding agencies. In this case, the emphasis was on strengthening the local communities' alternative land-use systems other than *jhum*, to take as much pressure as possible off *jhum* itself, so that it could operate at as long a fallow cycle as possible. Furthermore, realizing that the forest fallow phase has been greatly weakened under shortened *jhum* cycles, effort was made to strengthen the fallow phase through an accelerated succession by introducing appropriately identified early successional tree species, based on the communities' value systems. In such an effort, socially selected and ecologically significant keystone species have a key role to

**BIODIVERSITY MANAGEMENT AND
SUSTAINABLE LIVELIHOOD DEVELOPMENT**

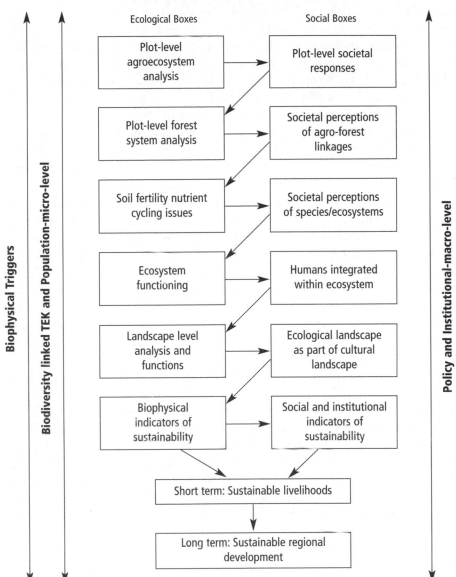

Figure 3.6 An iterative process for building linkages between ecological and social sciences, and
analyses of land-use dynamics from plot to landscape levels, leading to sustainable
livelihood/development of traditional societies living in biodiversity-rich areas

Source: Ramakrishnan (2001)

ensure community participation. In other words, the local communities' TEK formed the basis from the start, bringing in formal knowledge as much as desirable (Ramakrishnan, 1992a, 2001; Ramakrishnan et al, 2005). Such land-use redevelopment, based on fallow management, formed the basis for the developmental initiative in Nagaland, now well known as the Nagaland Environmental Protection and Development (NEPED) project taken up by the Nagaland Government, with funding through the India–Canada Environmental Facility (NEPED and IIRR, 1999). Thus, institution-level TEK was also seen as important as a result of research initiatives. It was therefore heartening to see that the Nagaland government gave much flexibility in developing the village development boards (VDBs), each ethnic group following their own ways of village-level institution building, but all having the same function – participation in rural development initiatives.

NEPED now involves all the villages of the State of Nagaland, and includes about 1200 villages and 200 experimental plots for agroforestry redevelopment, with replicated test plots covering about 5500ha. Farmers have adopted tree-based strengthened *jhum* systems based on agroforestry principles for local testing on about 33,000ha in 870 villages (38ha per village × 870 villages); in these plots, local adaptations and innovations for activities such as soil and water management are emphasized. Socially selected species such as Nepalese alder (*Alnus nepalensis*) were the key for successful adoption by local communities. In consultation with local communities, ten tree species selected for poles for house construction and fuelwood that could be harvested five to ten years after planting, and 20 tree species of value for timber, have been identified and introduced into *jhum* plots to strengthen the *jhum* system. Traditional rainwater harvesting systems and erosion control measures are incorporated into the redeveloped *jhum* practices, where appropriate. Community-based participatory biodiversity conservation (both crop- and forest-based) and carbon sequestration were important considerations, since tree species for fallow management were selected through a consultative process involving local communities. Land-use redevelopment emphasized participatory extension and dissemination, with gender considerations, through the VDBs based on the local value system.

These efforts have now led to renewed interest in the northeastern region, where many of the seven state governments are keen to initiate *jhum* redevelopment initiatives. The basic philosophy for development in all of these efforts has been to emphasize use of the concept of 'cultural landscape' as a window to link conservation with development, concurrently emphasizing diversification of developmental efforts through ecotourism or cultural tourism – and even to have some of the unique sites such as the Apatani cultural landscape inscribed as a natural cultural World Heritage Site by UNESCO. The outreach for these developmental efforts is again through both written and audio-visual media, along with location-specific research initiatives, building on more than three decades of experience.

References

Bhadauria, T. and Ramakrishnan, P. S. (1989) 'Earthworm population dynamics and contribution to nutrient cycling during cropping and fallow phases of shifting agriculture (*jhum*) in north-eastern India', *Journal of Applied Ecology*, vol 26, pp505–520

Bhadauria, T. and Ramakrishnan, P. S. (1991) 'Population dynamics of earthworms and their activity in forest ecosystems in north-east India', *Journal of Tropical Ecology*, vol 7, pp305–318

Bhadauria, T. and Ramakrishnan, P. S. (1996) 'Role of earthworms in nitrogen cycling during the cropping phase of shifting agriculture (*jhum*) in north-east India', *Biology and Fertility of Soils*, vol 22, pp350–354

Boojh, R. and Ramakrishnan, P. S. (1982a) 'Growth strategy of trees related to successional status: I. Architecture and extension growth', *Forest Ecology and Management*, vol 4, pp355–374

Boojh, R. and Ramakrishnan, P. S. (1982b) 'Growth strategy of trees related to successional status: II. Leaf dynamics', *Forest Ecology and Management*, vol 4, pp375–386

Elzinga, A. (1996) 'Some reflections on post-normal science', in Rolen, M. (ed.) *Culture, Perceptions and Environmental Problems: Interscientific Communication on Environmental Issue*, Stockholm, Swedish Council for Planning and Coordination of Research

FAO (1995) *Forest Resources Assessment 1990: Global Synthesis*, Forestry Paper 124, Rome, Food and Agriculture Organization

FAO/UNEP (1982) *Tropical Forest Resources*, Forestry Paper 50, Rome, Food and Agriculture Organization

Gangwar, A. K. and Ramakrishnan, P. S. (1987) 'Cropping and yield patterns under different land use systems of the Khasis at higher elevations of Meghalaya in north-eastern India', *International Journal of Ecology and Environmental Sciences*, vol 13, pp73–86

Gangwar, A. K. and Ramakrishnan, P. S (1989a) 'Cultivation and use of lesser-known plants of food value by tribals in north-east India', *Agriculture Ecosystem and Environment*, vol 25, pp253–267

Gangwar, A. K and Ramakrishnan, P. S. (1989b) 'Ecosystem function in a Khasi village of the desertified Cherrapunji area in north-east India', *Proceedings of the Indian Academy of Sciences (Plant Sciences)*, vol 99, pp199–210

Gangwar, A. K. and Ramakrishnan, P. S. (1990) 'Ethnobiological notes on some tribes of Arunachal Pradesh, North Eastern India', *Economic Botany*, vol 44, pp94–105

Gliessman, S. R. (1988) 'Ecology and management of weeds in traditional agroecosystems', in Altierie, M. A. and Liebman, M. (eds) *Weed Management in Agroecosystems: Ecological Approaches*, Boca Raton, Florida, CRC Press

Gliessman, S. R. (1990) 'Integrating trees into agriculture: The home garden agroecosystem as an example of agroforestry in the tropics', in Gliessman, S. R. (ed.) *Agroecology: Researching the Ecological Basis for Sustainable Development*, New York, Springer

Khiewtam, R. S. and Ramakrishnan, P. S. (1989) 'Socio-cultural studies of the sacred groves at Cherrapunji and adjoining areas in north-eastern India', *Man in India*, vol 69, pp64–71

Khiewtam, R. S. and Ramakrishan, P. S. (1993) 'Litter and fine root dynamics of relict sacred grove forest of Cherrapunji in north-eastern India', *Forest Ecology and Management*, vol 60, pp327–344

Kumar, A. and Ramakrishnan, P. S. (1990) 'Energy flow through an Apatani village ecosystem of Arunchal Pradesh in northeast India', *Human Ecology*, vol 18, pp315–336

Lambin, E. F., Turner II, B. L., Geist, H. J., Agbola, S., Angelsen, A., Bruce, J. W., Coomes, O., Dirzo, R., Fischer, G., Folke, C., George, P. S., Homewood, K., Imbernon, J., Leemans, R., Li, X., Moran, E. F., Mortimore, M., Ramakrishnan, P. S., Richards, J. F., Skånes, H., Steffen, W., Stone, G. D, Svedin, U., Veldkamp, T., Vogel, C. and Xu, J. (2001) 'The causes of land-use and land-cover change: Moving beyond the myths', *Global Environmental Change*, vol 11, pp261–269

Maikhuri, R. K. and Ramakrishnan, P. S. (1990) 'Ecological analysis of a cluster of villages emphasizing land use of different tribes in Meghalaya in north-east India', *Agriculture Ecosystem and Environment*, vol 31, pp17–27

Maikhuri, R. K. and Ramakrishnan, P. S. (1991) 'Comparative analysis of the village ecosystem function of different tribes living in the same area in Arunachal Pradesh in north-eastern India', *Agricultural Systems*, vol 35, pp377–399

Mishra, B. K. and Ramakrishnan, P. S. (1982) 'Energy flow through a village ecosystem with slash and burn agriculture in north-eastern India', *Agricultural Systems*, vol 9, pp57–72

Mishra, B. K. and Ramakrishnan, P. S. (1984) 'Nitrogen budget under rotational bush fallow agriculture (*jhum*) at higher elevations of Meghalaya in north-eastern India', *Plant and Soil*, vol 81, pp37–46

NEPED and IIRR (1999) *Building Upon Traditional Agriculture in Nagaland*, Nagaland, India and Silang, Philippines, Nagaland Environmental Protection and Economic Development and International Institute of Rural Reconstruction

Nye, P. H. and Greenland, D. J. (1960) *The Soil Under Shifting Cultivation*, Technical Communication No 51, Harpenden, Commonwealth Bureau of Soils

Ramakrishnan, P. S. (1986) 'Morphometric analysis of growth and architecture of tropical trees', in *Naturalia Monspeliensia: Colloque International Arbre*, pp209–222, Montpellier, France, Institut de Botanique

Ramakrishnan, P. S. (1989a) 'Nutrient cycling in forest fallows in north-eastern India' in Proctor, J. (ed.) *Mineral Nutrients in Tropical Forest and Savanna Ecosystems*, Oxford, Blackwell

Ramakrishnan, P. S. (1989b) 'Successional theory: Implication for weed management in shifting agriculture, mixed cropping and agroforestry systems', in Altieri, M. A. and Liebman, M. (eds) *Weed Management in Agroecosystems: Ecological Approaches*, Boca Raton, Florida, CRC Press

Ramakrishnan, P. S. (1991) *Biological Invasion in the Tropics*, New Delhi, National Institute of Ecology and International Science Publishers

Ramakrishnan, P. S. (1992a) *Shifting Agriculture and Sustainable Development: An Interdisciplinary Study from North-Eastern India*, Man and Biosphere Series, Book 10, Paris and Carnforth, UNESCO and Parthenon Publishing

Ramakrishnan, P. S. (1992b) 'Tropical forests: Exploitation, conservation and management', *Impact of Science on Society*, vol 42, no 166, pp149–162

Ramakrishnan, P. S. (1996) 'Conserving the sacred: From species to landscapes', *Nature and Resources*, vol 32, pp11–19

Ramakrishnan, P. S. (2001) *Ecology and Sustainable Development*, New Delhi, National Book Trust

Ramakrishnan, P. S. and Toky, O. P. (1981) 'Soil nutrient status of hill agro-ecosystems and recovery pattern after slash and burn agriculture (*jhum*) in north- eastern India', *Plant and Soil*, vol 60, pp41–64

Ramakrishnan, P. S. and Vitousek, P. M. (1989) 'Ecosystem-level processes and consequences of biological invasions', in: Drake, J. A. Mooney, H. A., di Castri, F., Groves, R. H., Kruger, F. J., Rejmanek, M. and Williamson, M. (eds) *Biological Invasions: A Global Perspective,* SCOPE Report 37, New York, John Wiley

Ramakrishnan, P. S., Saxena, K. G. and Chandrasekhara, U. (eds) (1998) *Conserving the Sacred for Biodiversity Management*, New Delhi and Oxford, UNESCO and IBH Publishers

Ramakrishnan, P. S., Boojh, R., Saxena, K. G, Chandrashekara, U. M., Depommier, D., Patnaik, S., Toky, O. P., Gangwar, A. K. and Gangwar, R. (2005) *One Sun, Two Worlds: An Ecological Journey*, New Delhi and Oxford, UNESCO and IBH Publishers

Rao, K. S. and Ramakrishnan, P. S. (1989) 'Role of bamboos in nutrient conservation during secondary succession following slash and burn agriculture (*jhum*) in north-eastern India', *Journal of Applied Ecology*, vol 26, pp625–633

Rossler, M. (2001) 'Sacred landscapes: New perspectives in the implementation of the cultural landscape concept in the framework of the UNESCO world heritage convention', in *UNESCO Thematic Expert Meeting on Asia-Pacific Sacred Mountains, 5–10 September 2001, Wakayama City, Japan*, Paris, World Heritage Centre, UNESCO, pp27–41

Ruthenberg, H. (1971) *Farming Systems in the Tropics*, Oxford, Clarendon Press

Saxena, K. G. and Ramakrishnan, P. S. (1984) 'Herbaceous vegetation development and weed potential in slash and burn agriculture (*jhum*) in north-eastern India', *Weed Research*, vol 24, pp135–142

Shukla, R. P. and Ramakrishnan, P. S. (1984a) 'Biomass allocation strategies and productivity of tropical trees related to successional status', *Forest Ecology and Management*, vol 9, pp315–324

Shukla, R. P. and Ramakrishnan, P. S. (1984b) 'Leaf dynamics of tropical trees related to successional status', *New Phytologist*, vol 97, pp697–706

Shukla, R. P. and Ramakrishnan, P. S. (1986) 'Architecture and growth strategies of tropical trees in relation to successional status', *Journal of Ecology*, vol 74, pp33–46

Spencer, J. E. (1966) *Shifting Cultivation in South-eastern Asia*, Berkeley, University of California Press

Swamy, P. S and Ramakrishnan, P. S. (1987) 'Weed potential of *Mikania micrantha* H. B. K. and its control in secondary successional environment after shifting agriculture (*jhum*) in north-eastern India', *Agriculture Ecosystem and Environment*, vol 18, pp195–204

Swamy, P. S and Ramakrishnan, P. S. (1988) 'Nutrient budget under slash and burn agriculture (*jhum*) with different weeding regimes in north-eastern India', *Acta Oecologia: Oecologica Applicata*, vol 9, pp85–102

Swift, M. J., Vandermeer, J., Ramakrishnan, P. S., Ong, C. K., Anderson, J. M. and Hawkins, B. (1996) 'Biodiversity and agroecosystem function', in Mooney, H. A., Cushman, J. H., Medina, E., Sala, O. E. and Schulz, E.-D. (eds) *Functional Roles of Biodiversity: A Global Perspective*, Chichester, John Wiley

Toky, O. P. and Ramakrishnan, P. S. (1981) 'Run-off and infiltration losses related to shifting agriculture (*jhum*) in north-eastern India', *Environmental Conservation*, vol 8, pp313–321

Toky, O. P. and Ramakrishnan, P. S. (1983a) 'Secondary succession following slash and burn agriculture in north-eastern India: I. Biomass, litterfall and productivity', *Journal of Ecology*, vol 71, pp737–745

Toky, O. P. and Ramakrishnan, P. S. (1983b) 'Secondary succession following slash and burn agriculture in north-eastern India: II. Nutrient cycling', *Journal of Ecology*, vol 71, pp747–57

UNESCO (1983) *Swidden Cultivation in Asia: Vol. 2, Country Profiles*, Bangkok, UNESCO Office of the Regional Adviser for Asia and the Pacific

World Commission on Environment and Development (1987) *Our Common Future*, Oxford, Oxford University Press

4

Policy-Oriented Conservation Design

David J. Mattson and Troy Merrill

The challenges of conservation design

Human populations have increased exponentially over the last three centuries and have placed exponentially increasing demands on nature, resulting in degraded ecosystems, extirpated biota and species extinctions. As human populations continue to increase, demands on energy sources, fresh water, arable land and natural sources of protein are expected to increase apace as human-caused atmospheric pollution additionally stresses biological systems, forced primarily through climate change. Organisms that depend on wilderness conditions for survival, that are adapted to a limited range of habitats or foods, or that are otherwise restricted to regions transformed by urbanization and agriculture have been and will continue to be the most vulnerable (Ceballos and Brown, 1995; Collar et al, 1997; Czech et al, 2000; Hendee and Mattson, 2002), with population fragmentation exacerbating vulnerabilities and foreshadowing extirpation (Debinski and Holt, 2000; Saunders et al, 1991).

Conservation goals and strategies have evolved in response to human impacts on biological systems during the last 200 years. Conservation and management of game populations were a hallmark of the 19th century, in response to widespread extirpations of game animals (Reiger, 2001). National parks and similar protected areas became a favoured prescription during the late 19th century, and remained so throughout the 20th century, primarily to protect scenic attractions and secondarily to preserve biological resources (Wright and Mattson, 1996). The idea of 'wilderness' and interest in preserving wilderness conditions evolved as a major rationale for protecting landscapes and associated biota during the 20th century, culminating in the 1970s and 1980s (Hendee and Dawson, 2002). Biological diversity and healthy or sustainable ecosystems, as such, became a focus of attention only during the late 20th century, as awareness of species losses and ecosystem degradation became more widespread (Noss and Cooperrider, 1994). Increasingly, conservation is equated with preserving biodiversity and sustaining healthy ecosystems (Groves, 2003; Noss and Cooperrider, 1994) – goals that have proven difficult and elusive.

Social and decision processes

Human values, world-views and myths are central to achieving biodiversity conservation because they largely shape what people expect, desire and demand from the world. Those who promote biodiversity conservation and the sustenance of healthy ecosystems tend to give comparatively greater weight to outcomes and effects relatively distant in time and space (Portney and Weyant, 1999; Thiele, 1999) and have a sense of rectitude shaped by biocentric or ecocentric myth systems (Manning et al, 1999). They also tend to prioritize physical and emotional well-being and, as a derivative of biocentric attitudes, experience respect and affection for wildlife and wild places (Kellert, 1989). By contrast, those who are indifferent towards, or unsupportive of, conservation goals tend to prioritize more narrowly defined and immediate considerations focused on power and wealth, or exhibit rectitude rooted in anthropocentric myth systems that justify the domination of nature and valuation based primarily on utility (Cawley, 1993; Kellert et al, 1996; Manning et al, 1999; Reading and Kellert, 1993; Reading et al, 1994). To craft and implement effective conservation policies, these diverse and potentially conflicting myths and value orientations must be considered because they largely determine how people define, talk about and respond to 'problems'.

The structure of societal decision processes largely determines whether or not diverse human demands can be reconciled in effective and sustainable conservation policies. In the interior west of Canada and the US, current power arrangements potentially preclude equitable and democratic decision-making, which is often a prerequisite for developing durable conservation policies and practices. Natural resource decision processes in this region have deferred to anthropocentric attitudes since European settlement (for example, Clark, 1997; Clark and McCool, 1996; Culhane, 1981; Lichatowich, 1999; Miller et al, 1996; Reisner, 1993). League of Conservation Voters (LCV) score cards show that virtually all of the contemporary elected officials in this region are ambivalent towards or unsupportive of conservation goals (www.lcv.org/scorecard/).

In common with tendencies of governance everywhere (Phillips, 2002; Solimano, 1998; Somit and Peterson, 1997), decision processes in this region also have a legacy of deference to power and wealth values reckoned over short time-frames (for example, Lichatowich, 1999; Reisner, 1993; Taylor, 1999; Wagner et al, 1995; Wilkinson, 1998; Yaffee, 1994). This regional tendency for natural resource decision processes to serve anthropocentric power and wealth interests has recently been curbed primarily by national-level policies that reflect a politically strong national, rather than regional, constituency for conservation goals (Keiter and Locke, 1996; Parker, 1995; Yaffee, 1994). The juxtaposition of national biocentric policies with regional anthropocentric preference among elites in natural resource decision processes has created conflict (for example, Cawley, 1993; Yaffee, 1994) heightened by a demographically driven trend towards more conservation-friendly attitudes among regional non-elites (Hansen et al, 2002; Smutny and Takahashi, 1999). This conflict has predictably resulted in polarization, widespread value deprivations, and erosion of common interests (Brunner, 2002).

The health and even survival of democratic governments depend on the prevalence of common over particularist interests in policy processes (Brunner, 2002; Clark, 2002). Democracy is at risk to the extent that polities fail to identify and secure their common interests, as is often signalled by widespread value deprivations (Brunner, 2002; Lasswell and McDougal, 1992). Under such conditions, no system of government can persist for long. Much like 'healthy' ecosystems, a 'healthy' human society is indicated by sustainable structure (equitable distribution of values) and sustainable function (processes that promote common interests) (Lasswell and McDougal, 1992). There is increasing evidence as well as belief that healthy human societies and healthy ecosystems are interconnected (Bromley and Paavola, 2002; Homer-Dixon, 1999; Thiele, 1999). Both are required for long-term durability and resilience. Any system of conservation design and management that aspires to create durable conservation solutions must take governance into consideration, which means paying serious attention to decision processes and people's diverse values, expectations and demands.

The limits of deterministic conservation design and adaptive management

The biophysical and human realms within which conservation practice occurs are almost always complex (Clark, 2002). Either domain considered alone is overwhelming in this regard, much less in combination. Considering biodiversity as a conservation focus, we have not yet even come up with an efficient means of measuring this quantity (Hawksworth, 1995). At best, we can track the status or loss of large or otherwise charismatic species that are a focus of societal attention and resources. Even considering well-studied species such as grizzly bears (*Ursus arctos horribilis*) or spotted owls (*Strix occidentalis*), complexities quickly outstrip both the cognitive abilities of individuals and the more formalized analytical and rule-making abilities of management agencies (Clark, 2000; Yaffee, 1994). However, when we look to the future, the situation is even worse. Current uncertainties entailed by existing complexity increase rapidly (Ludwig et al, 1993). Exponentiating incertitude is guaranteed by climate change alone (Karl and Trenberth, 2003), even without considering other global trends. Confident predictions about any phenomena of even modest complexity have been and will continue to be elusive (Kravtsov, 1993).

A deeply rooted human tendency to discount future events compounds the effects of uncertainty insofar as human decision processes are concerned (Heinen and Low, 1992; Miles, 1998). Although varying somewhat with age, economic status and culture, humans exponentially discount temporally distant rewards or penalties compared to those nearer at hand (Puttaswamaiah, 2002). As a consequence, the intrinsic uncertainty of the future is readily invoked as cause to not consider even foreseeable events in current decision-making, especially when status quo power and wealth arrangements are at stake (Dasgupta, 2001; Portney and Weyant, 1999). This raises the question of how people of good intentions can address future uncertainties.

Especially during the last 20 years, adaptive management has been the most common prescriptive response to the dual considerations of uncertainty and potentially weighty future events (Lee, 1993): 'we will manage adaptively'. This usually means that prescriptions are treated as hypotheses to be tested through application and appraised based on monitoring or other ongoing investigation (Lee, 1993). If outcomes deviate from predictions, prescriptions are modified according to contemporary understanding of the system of interest. Of relevance to democratic decision-making, applications of adaptive management have largely focused on biophysical elements of managed systems and not on elements pertaining to governance, culture and human psychosociology. To be sure, adaptive management can, in itself, be a framework for facilitating common interest decision processes (Lee, 1993). However, this is often ancillary or implicit rather than focal.

Adaptive management has great appeal. At the very least, it provides a ready answer by those in power to others concerned about momentous future events such as extinction. However, even with sincere applications, there are potentially major problems related primarily to the nature of natural resource decision processes and bureaucratic organizations. In particular, adaptive management presupposes the existence of natural resource management agencies that learn well and embrace new knowledge about somewhat ill-defined resources such as biodiversity, as well as new knowledge about the efficacy of existing practices and related institutional arrangements (Lee, 1993; Miller, 1999). Unfortunately, such entities may be rare. In fact, there is ample evidence for how poorly bureaucratic organizations generally learn, and how unresponsive they are to information that calls encultured or otherwise institutionalized practices into question (for example, Argyris and Schön, 1978; Clark, 1993; Gareth, 1986; Miller, 1999). More to the point, even if adaptive management were to perform precisely as designed, bureaucratic agencies typically respond to information to the extent that there is a political or policy imperative to do so (for example, Ingram, 1973; Sabatier, 1978; Wilson, 1989). Clearly, adaptive conservation design and adaptive policy processes have a place in dealing with the intrinsic uncertainties of complex systems and with an even more uncertain future. The question is, what does an adequate adaptive process look like and what is the focus of attention?

Tool-oriented conservation design: Reserves, co-management and other approaches

Most established and named approaches to conservation design can be characterized as 'tool-oriented'. That is, they tend to define problems and related solutions in the terms of the tool or approach that is being promoted rather than in terms of the content and context of a given situation. In the vernacular, given a 'hammer', all problems become a 'nail'. Insufficient attention is often given to clarifying the goals of participants in a specific case, mapping problematic trends and conditions, creating alternative solutions, understanding the social and political context, and locating the current situation and potential solutions within relevant policy processes. Often the approaches become politicized, in the sense that attention is focused on who is exhibiting conceptual purity or fidelity in application and who, then, warrants receiving approbation and resources.

The current paradigm of designing reserve or protected area networks exemplifies a tool-oriented approach to conservation problem-solving. Reserve or conservation area design (CAD) is largely a matter of determining the size, shape and location of reserves invoking the concepts of cores, buffers and corridors (Groves, 2003; Noss, 2003; Noss and Cooperrider, 1994; Noss et al, 1997). Cores identify 'protected' areas where human impacts are minimal; buffers identify surrounding lands with multiple human activities managed so as to be compatible with conservation goals; and corridors are areas that connect and allow for movement of organisms between cores. This is all in concept. In reality, 'protected' areas are the outcome of diverse prescriptions that allow any number and types of human impacts. Some protected areas allow hunting, others allow grazing and most allow commercial development. Moreover, whatever the prescription, realized protection depends on resources and political will, to the point where some 'protected' areas receive no protection at all (see for example, Aiken, 1994; Barany et al, 2002; De Lopez, 2001; Inamdar et al, 1999; Lusigi, 1994). Even if reserves are adopted as the primary prescriptive response to conservation problems, it quickly becomes clear that effective conservation design also requires coming to grips with the complexities of social and decision processes. More important, 'reserves' are often not the appropriate prescription for a conservation problem (Salafsky et al, 2002; Wilshusen et al, 2002). A formulaic or tool-based approach based on reserves is not sufficient and may even be detrimental.

A number of approaches to conservation design fall outside the paradigm of reserve design or CAD and emphasize additional considerations or alternate prescriptions (Redford et al, 2003). However, they are still tool-based or *ad hoc* in approach. For example, there is a large body of thought and practice organized around community-based or collaborative conservation (for example, Brosius et al, 1998; Steelman and Ascher, 1997; Wondolleck and Yaffee, 2000). The focus is on social process and finding common ground among diverse participants as a means of achieving local conservation goals. Although often uncritically promoted (for example, Wondolleck and Yaffee, 2000), there is ample evidence that the success even of this attractive approach to conservation problem-solving is highly contingent on factors such as transience of participants, opportunities to defect, number of issues and available information (Fukuyama, 2000; Singleton, 2000; Weber, 1998; Wondolleck and Yaffee, 2000). Context and content matter. There is also a great deal of conservation design motivated by formal conservation policies including, in the US, the Endangered Species Act (16 US Code §§ 1531–1544), the National Environmental Policy Act (42 US Code §§ 4321–4347), and the 'organic' act of the National Park Service (16 US Code §§ 1–4). These formal design processes take into consideration various social and biological factors, but they are all quite rigid and context-insensitive by intent (for example, Lachapelle et al, 2003; Lee, 1993; Miller et al, 1994; Yaffee, 1994). Context is relevant only to the extent of determining where these formal policies are to be applied. Finally, there are innumerable *ad hoc* conservation designs that have arisen in response to the exigencies of specific cases and that are highly contingent on local human resources and the idiosyncratic ways in which the case was conceptualized. None of these approaches are intended to provide a frame work that is explicitly problem-oriented, sensitive to the biological and *human* context, and designed to maximize rationality. What is needed is an approach that does all of this.

Policy-oriented conservation design

At its most fundamental, conservation is about stopping destructive human behaviours or, conversely, promoting and preserving behaviours that are beneficial. This follows from the fact that most causes of ecological damage and species endangerment are rooted in widespread human attitudes and practices (Czech et al, 2000; Kerr and Currie, 1995; McKinney, 2001). Moreover, as much as human behaviours are the root of most conservation problems, they are also the source of most solutions (Ehrlich, 2002; Heinen and Low, 1992). Conservation design is, in essence, the articulation of goals and the means of achieving them, defined primarily in terms of decisions needed to instigate, modify or preserve certain human behaviours in certain places at certain times (Salafsky et al, 2002). Human decision-making and policy processes are central to the business (Clark, 2002).

The demands on conservation designers are great, especially if they aspire to be effective. They need to understand human social psychology (Miller, 1999; Oskamp, 1995) and be expert at operating in complex policy systems (Clark, 2002). They also need to have enough insight into ecological systems to specify what 'sustainable' ecosystem structure and function look like at any given place and time. Moreover, given that every conservation organization has limited resources, they must have enough insight into ecological and policy processes to allow for the strategic allocation of resources across multiple temporal and spatial scales. Under such circumstances, being effective requires an adaptive framework that efficiently guides inquiry and the promotion, development, implementation and appraisal of policies, at the scale of both organizations and societies (Clark, 2002). In what follows, we describe one possible framework for organizing inquiry related to conservation design, provide examples of its partial application, and discuss some relevant context. The approach, which we call policy-oriented conservation design, is an adaptation of the intelligence and prescription functions of the policy sciences (Clark, 2002).

The concept of policy-oriented conservation design

As implied by the name, policy-oriented conservation design (POCD) focuses on decision and policy processes. We use the term 'policy' here to mean any understanding or agreement, formal or informal, arrived at by a collective of people regarding what should be done about a matter of common interest (Clark, 2002). Usually these 'matters' pertain to the allocation of resources or opportunities. In this sense, policy encompasses most human activities related to governance and collective decision-making. It is also about politics, or who gets what, when and how (Lasswell, 1950). Decision-making is integral to the policy process (Clark, 2002). Decisions are at the core of prescribing, implementing and appraising policy. Decision-making is also at the core of human practices and behaviours that can either cause conservation problems or contribute to conservation solutions. Essentially, successful conservation depends on individual people *deciding* to create policies or engage in behaviours that promote rather than degrade sustainable ecosystems.

By its nature, POCD is problem-oriented. That is, POCD focuses on solving problems. In this sense, a problem is any discrepancy between where we are and where we want to be (Dery, 1984). The objective is to bridge this gap. Given the complexity of most situations, successful bridging requires creativity, multiple methods, diverse perspectives and critical analysis (Clark, 2002). Hidebound adherence to formulaic prescriptions is antithetical. As commonsensical as this may seem, most people find it very difficult to undertake truly problem-oriented conservation design. Disciplinary bias, rules of thumb, simple formulas and the 'trap of experience' are commonplace (see for example, Clark, 1993; Edmunds, 1987; Gowda, 1999; Hastie and Dawes, 2001; Mellers et al, 1998). As much as we are told by the Myth of the Enlightenment that humans are rational in some transcendent sense, there is ample evidence that bounded rationality and the acontextual use of simple mental models are the norm in human affairs (Gigerenzer and Goldstein, 1996; Markman and Gentner, 2001). For this reason, POCD can rarely be done well on an *ad hoc* basis, or on the basis of good intentions alone, but requires some kind of framework that promotes rationality and problem-solving.

POCD operates on the premise that prescription is highly contextual. In other words, one's plans depend entirely on situation-specific goals, trends, conditions and projections, for both the biophysical and human realms. The human realm can be characterized by the identities, expectations and demands of human participants, as well as by the values that participants hold, the strategies they use to interact, the outcomes of their interactions and the effects that transpire (Clark, 2002). By this approach, the appropriate prescription, or design, depends entirely on the situation. Moreover, the quality of a given design depends on the creativity of those involved and their ability to efficiently analyse the relevant human and biophysical systems. To the extent it can be attained, rationality is important. In this sense, 'rationality' means achieving as clear an understanding as possible of relevant factors and relevant interactions, transcending ego defences and all of the blinkers (blinders) introduced by simple rubrics built on disciplinary training and circumscribed experience. A somewhat unconventional view of science and analysis is required to achieve this end.

The intelligence function in POCD

The collection, analysis and promulgation of information in POCD are best described as intelligence activities. 'Intelligence' here is used in the sense of military intelligence, or the application of information to achieve success. Intelligence activities can be grouped under the intelligence function of the policy process (McDougal et al, 1981). The intelligence function includes both 'science' and 'analysis' (Brewer, 1981; Clark, 1992). Science aims to produce generalizable collective knowledge. Analysis, at least in the sense used here, is devoted to gaining situation-specific insight that will facilitate success. The objective is not necessarily to contribute to a collective pool of knowledge; insight or knowledge may be held only by those involved in developing a design or otherwise solving a problem.

Because the intelligence function is intrinsically problem-oriented, its associated activities are subject to standards different from those conventionally applied to science (McDougal et al, 1981). The primary standard for the intelligence function is 'sufficiency' (Ravetz, 1996). In other words, are the available knowledge and insight sufficient for those involved to achieve their specified goals? More specifically, is the intelligence function sufficiently practical and informative (Clark, 2000; Lasswell, 1971)? Practicality refers to whether information is being generated in a timely, accessible and cost-effective manner. Informative refers to whether the information is comprehensive, relevant and 'reliable'. In almost all undertakings, there is limited time and money. Information needs to be affordable and available 'in time'. Information also has to be relevant to the issues at hand, comprehensive enough to deal with all of the important identified factors and relationships, and as congruent as possible with 'the data' or other empirical bases. Employing this paradigm, people involved in intelligence activities are not involved in the pursuit of 'truth' to the exclusion of all other considerations. Rather, they are involved in producing information and insight that is comprehensive, relevant and reliable enough to solve the problem or otherwise achieve success, and do so in a timely and cost-effective manner. As in all other aspects of POCD, the sufficiency of the intelligence function is highly context-specific. The information sufficient to meet needs in one context may not be sufficient to meet needs in another, depending on available resources, the biophysical system and situation-specific human, social and decision processes. And, clearly, sufficiency is wholly contingent on goals.

Many conservation endeavours aspire to being 'science-based' – for example, the Predator Conservation Alliance (www.predatorconservation.org) and the Yellowstone-to-Yukon (Y2Y) Conservation Initiative (www.y2y.net). This raises the question: how does science-based relate to sufficient intelligence? In most instances, the term 'science-based' is used to mean employing 'reliable' information about biophysical elements in conservation planning (Noss, 2003). As an important aside, 'reliable' information about the human context is rarely considered. Traditional standards of science are also invoked for determining whether information is 'reliable' or not, for example, by withstanding critical tests. Great emphasis is placed on minimizing the likelihood of concluding that some effect or relationship exists, when it does not (that is, avoiding type I errors).

There have been many compelling critiques of the naïve practice of science as focused on 'reliability' and minimizing type I errors, especially in conservation applications (for example, Formaini, 1990; Ludwig et al, 1993; Shrader-Frechette and McCoy, 1993). There is not the space here to review them. Suffice it to say that, in many instances, it is more important to avoid concluding that a relationship or effect does not exist, when it does (that is, avoiding type II errors) (Shrader-Frechette, 1991). It is also often more relevant to know what the *most defensible* representation of a biological or human system is, rather than whether some hypothesis or theory has withstood a number of critical tests (Lemons, 1996). Those who are involved in conservation design need to know which, of all the possible ways they can understand how the world works, is most likely to produce success. Moreover, they need to make this judgement regarding *all* of the factors likely to affect the success of conservation endeavours, and not just regarding factors that have been subject to exhaustive scientific investigation. In short, we do not find 'science-based', as conventionally construed, to be a very useful concept for guiding conservation design and problem-solving. Because of the conceptual and semantic baggage attached to 'science-based', we prefer to employ the concept of 'sufficient intelligence' in POCD.

An approach to POCD

The approach to POCD that we describe here is, in some sense, formulaic.

Formulas are inescapable – a derivative of values, world-views and mental models (Hastie and Dawes, 2001; Shafir and LeBoeuf, 2002). However, POCD is *not* intended to be formulaic in the sense of a blanket policy prescription or specific tool. Rather, it is formulaic at the level of analysis. This is an important distinction given that our goal is to enhance both rationality and appreciation of context with a focus on solving problems rather than promoting dogma. Formulas that go so far as to explicitly or implicitly prescribe policies and practices without regard for site-specific conditions bypass context and discount rationality.

POCD ideally consists of five main elements: first, clarification of the problem-solver's standpoint; second, orientation to the 'problem'; third, 'mapping' of the context; fourth, specification of a design, policy or approach; and fifth, subsequent appraisal and redesign. This process can also be understood as figuring out what is wanted and why, where to focus and what to do there, how to get change where and when it is wanted, and learning.

1   *Standpoint clarification* entails diagnosing one's own identity and, from that, reaching some understanding of one's biases and limitations (Clark and Wallace, 2002a). This insight can be taken into account during other parts of the design or problem-solving process. This is a critical and often overlooked step.
2   *Problem orientation* entails clarifying goals, identifying key factors governing goal-related outcomes, determining past trends in key factors, projecting trends into the future and developing alternatives for action (Clark, 2002). Problem orientation helps clarify what is wanted, the 'things' relevant to assessing and specifying outcomes, and what has happened in the past and is likely to happen in the future with factors that define the problem.
3   *Context mapping* includes describing and analysing conditions in both the biophysical and human realms (Machlis et al, 1997) that are relevant to specifying and achieving outcomes. Methods for assessing relevant structure and function of the biophysical realm have been described in detail (for example, Groves, 2003; Noss, 2003), although we recommend applying these methods with appropriate sensitivity to context. As briefly described above, methods for specifying relevant features of the human realm include identifying participants and characterizing their identities, expectations and demands. The situations that participants operate in, the strategies they use, the outcomes they achieve, and, ultimately, effects on societal 'structure and function' are also ideally described. A key part of each person's identity is the values they prioritize and the key symbols and 'narratives' that they adopt for explaining the world to themselves and others (see above). Much of this information can be presented in the form of conceptual models that are figurative maps of social process (Clark, 2002; Clark and Wallace, 2002b).

4   Drawing on information from the other steps, a *design* specifies desired outcomes, either or both on the ground and in social and decision processes, and also provides guidance regarding actions that need to be taken, and by whom, to achieve these ends. Actions are specified in terms of different facets of policy and decision processes, whether it be promoting and prescribing new policies, appraising and improving the implementation of existing ones, or terminating policies that are detrimental (Clark, 2002). Again, policies can either be formal, as in written laws, or informal, as in conventional practices.

5   Finally, *appraisal and redesign* allow for the adaptive consideration of new information on performance and context. Appraisal could be construed, in part, as the monitoring phase of adaptive management. However, this facet of POCD is intended to be more comprehensive, in that it sets standards and considers responses in both the biophysical and human systems, including the agency or organization motivating the design process.

The chances of producing or perpetuating an insufficient design increase if any facet of the design process is neglected altogether. However, the level of detail or rigour and the corresponding investment of time and energy required for each element depend, as always, on the context.

Example: Elements of POCD in the case of Rocky Mountain grizzly bears

In the Appendix to this chapter, we provide an outline that was developed to provide direction for some facets of POCD in a specific case. This example emphasizes: clarifying goals; translating goals into focal elements or biophysical features; identifying related stressors; identifying associated participants, practices and policies; determining what practices or policies need to be changed in what ways, and by whom; and frequent appraisal. It is a limited example only and is not appropriate for many situations. Its value is primarily in showing a translation of the rather abstract notions described above into terms more concretely applied to a specific conservation endeavour.

'Preservation' and 'restoration' planning areas were differentiated for purposes of this particular application. Preservation planning areas are places where conditions are currently acceptable and efforts are focused on planning to keep things the way they are. Restoration planning areas are places where conditions deviate from goals, and efforts are focused on planning for improvement. The terms 'reserves' or 'cores' were not used because the prescription of legislated protected areas was not necessarily an intended outcome. Rather, the intent was to differentiate planning and design depending upon whether improvement or preserving the *status quo* was emphasized. On another qualifying note, this example was premised on goals being rooted in biophysical conditions. In other words, the focus was on determining goals and desired outcomes explicitly for the biophysical realm, with all else deriving from this consideration. For reasons that we discuss in the concluding section of this chapter, such a premise may not be appropriate for many complex conservation cases.

We undertook an analysis focused on grizzly bear conservation in the North American Rocky Mountains that illustrates some elements of POCD. The overarching goal motivating our analysis was the establishment of evolutionarily robust populations of grizzly bears (see below) throughout the Rocky Mountains. Our analysis invoked the assumption that appraisals of biophysical conditions should frame or otherwise preconfigure analysis of social and decision processes, recognizing that, under many circumstances, this assumption may complicate rather than promote the achievement of conservation goals. Our analysis objectives were to:

1 identify preservation and restoration planning areas based on reckonings of grizzly bear population viability;
2 analyse factors limiting grizzly bear populations in each area, including trends, current conditions and future projections for each;
3 similarly analyse trends, conditions and projections for local social and decision processes; and
4 based on the second and third objectives, develop area-specific strategic options.

Given these analysis objectives, our approach was multi-scale and hierarchical. We first determined the extent and location of habitat able to support growing populations of grizzly bears ('source' areas), then determined whether each existing or potential population was demographically and evolutionarily robust, and then identified regions critical to restoring the demographic or evolutionary viability of source areas that were below the optimum. Focusing in on key areas, we then used predictive models or local grizzly bear data to determine critical factors ('stressors') limiting habitat capability. We also developed social process maps and other analytical summaries to gain insight into opportunities for promoting, prescribing or implementing gainful human practices, or ending human practices that were detrimental to grizzly bear conservation. We then generated strategic options that we scored for relative effectiveness in restoring grizzly bear habitat and for relative 'doability' or tractability based on social and political considerations – that is, roughly corresponding to the standards of reasonable and possible (Clark, 2000). Options judged to be the most efficacious and tractable were recommended for selection.

We used a model from the Yellowstone region of Montana, Wyoming and Idaho for our broad-scale appraisal of potential source areas (Merrill and Mattson, 2003). The model was based on distributions of dead grizzly bears in this ecosystem, and identified human population size, density of road and trail access, habitat productivity, national park protection and location in or out of a public grazing allotment as the primary predictors of where grizzly bears died. We generalized model results to the scale of a typical female grizzly bear life range, so as to represent landscape conditions at the scale grizzly bears lived and died. After identifying potential population source areas, we used another model to determine the probable number of bears that could be supported within each (Merrill, 2004). Based on this result, we then judged whether populations identified with each source area were likely to be demographically and evolutionarily robust, that is, likely to survive decadal environmental variation, and, accounting for genetic considerations, millennial environmental variation as well. Based on previous empirical and theoretical work, we judged populations of >500–700 and >2000 animals, respectively, to be demographically and evolutionarily robust (Merrill, 2004).

Several key results arose from the broad-scale appraisal (see Figure 4.1). Our modelling suggested that none of the potential source areas in the contiguous US could support an evolutionarily robust population (ERP). Consistent with previous modelling results (Boyce and Waller, 2003; Mattson and Merrill, 2002; Merrill et al, 1999), we determined that currently unoccupied habitat in central Idaho was capable of supporting perhaps the largest demographically robust population (DRP) south of the Canadian border, and was potentially a critical connection between the Yellowstone grizzly bear population and the ERP in Canada. Given the objective of creating an ERP south of Canada, we also identified two critical zones of fragmentation, the first separating the occupied range in the Yellowstone region from the potential range in central Idaho, and the second separating the potential range in central Idaho from the ERP in Canada. The second zone of fragmentation, centred on the Cabinet-Yaak region of northwestern Montana and adjacent Idaho and British Columbia, was of particular interest because it contained remnant grizzly bear populations. For that reason, we analysed biophysical and socio-political conditions in this region to craft restoration alternatives.

We developed another source area model for the Cabinet-Yaak region, based on local grizzly bear data (Mattson and Merrill, 2004). This model identified human population density as paramount in determining local grizzly bear densities, followed by density of road access. Our modelling also suggested that human lethality, rooted in human attitudes and practices, was as important a factor as human density *per se*. Our results showed that if current rates of human population growth continued, grizzly bear source areas would virtually disappear in this region (see Figure 4.2). From this, we concluded that regional conservation strategies should consider not only road density, but also human population distribution and growth, and human attitudes and practices that result in dead bears.

We consulted several individuals familiar with regional social and political processes to develop a social process map for the Cabinet-Yaak region (see Figure 4.3) and to identify opportunities for developing common ground. This

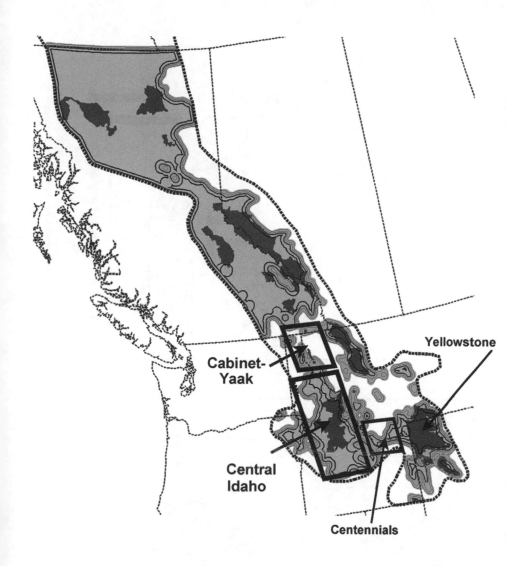

Figure 4.1 Areas within the North American Rocky Mountains (dashed line) where grizzly bear populations are predicted to be potentially stable or growing (source areas)

Note: Source areas are shaded grey; the darkest grey denotes protected areas (for example, national parks and wilderness areas). Source area size depends on different assumptions about human lethality (the probability that a human will kill a grizzly bear upon encountering one). Source areas increase progressively in size, as shown by concentric lines, if humans are assumed to be less lethal. The currently unoccupied central Idaho range is identified along with important zones of fragmentation separating the Yellowstone region from central Idaho (the Centennials) and central Idaho from grizzly bear populations in Canada (the Cabinet-Yaak).

Source: Based on models in Merrill and Mattson (2003)

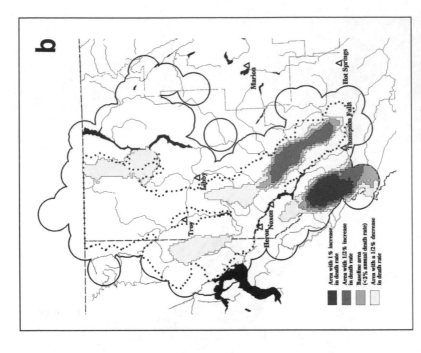

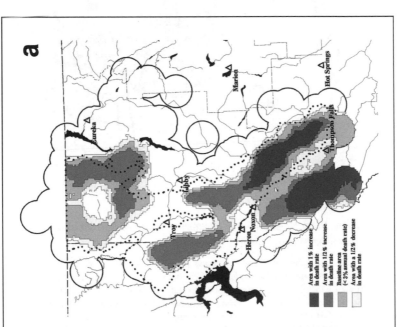

Figure 4.2 Areas within the Cabinet-Yaak region (outer solid line) where grizzly bear populations are predicted to be potentially stable or growing

Note: Predictions are based on (a) current conditions and (b) a 1.5-fold increase in regional human population sizes. Source areas are grey, with various shadings corresponding to changes in spatial extent associated with different assumptions about human lethality and related changes in annual grizzly bear death rates (that is, 0.5–1 per cent increase or 0.5 per cent decrease over baseline conditions). Darker grey indicates higher probability of being a source area.

Source: Based on models in Mattson and Merrill (2004)

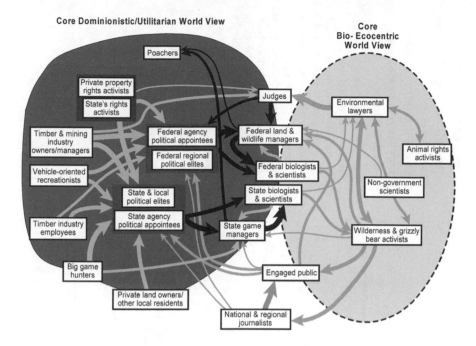

Figure 4.3 A map of social process for grizzly bear conservation
in the Cabinet-Yaak region

Note: The domains of prevailing world-views and associated myth systems (dominionistic/utilitarian or
anthropocentric; and biocentric or ecocentric) are shown. Generic participants are identified in boxes.
Dominant interactions and effects among participants are shown by arrows. Those that are predominantly coercive are black; those that are predominantly persuasive are grey. The width of arrows is
roughly correlated with the estimated strength of effects.

entailed developing profiles for generic participants that consisted of prioritized
values and value demands, prevalent myths and narratives, key relations, the
relative strength of relations, and whether relations were potentially coercive or
entirely based on persuasion. Local and regional elected and appointed political leaders were almost wholly identified with utilitarian and dominionistic
(anthropocentric) myths and narratives (Gagnon Thompson and Barton, 1994;
Kellert, 1989) shared with regional commodity interest groups. On this basis, as
well as by control over access to power and wealth, special interests organized
around commodities had almost exclusive access to elected and appointed
political officials through persuasion. Such officials, in turn, had recourse to
powerful means of promoting commodity special interests through sanctioned
policy and decision processes (for example, by control of agency budgets). By
contrast, participants promoting grizzly bear conservation and identified with
ecologistic, moralistic or humanistic (ecocentric) myths (Gagnon Thompson

Figure 4.4 Plots showing the estimated effectiveness ('Effect') and tractability of seven potential strategic directions for grizzly bear conservation in the Cabinet-Yaak region

Note: The different symbols represent the judgement of three different experts. The ellipse surrounding each cluster of symbols indicates the collective uncertainty regarding effectiveness and tractability for each potential strategic direction. Prioritized strategies were located in the upper right-hand quadrant (effective and tractable). By contrast, although human population growth was judged to have a major effect on grizzly bears in the Cabinet-Yaak region, it was also judged to be an intractable issue.

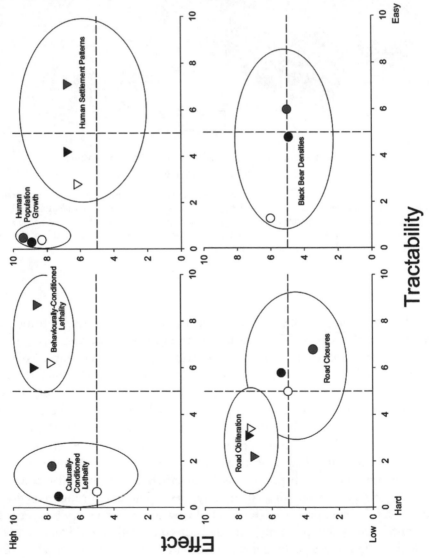

and Barton, 1994; Kellert, 1989) had virtually no access to regional decision processes other than by weak persuasion. Litigation by application of national environmental policies was the only exception. However, this coercive mechanism created destructive antagonism between local conservationists and natural resources agency decision-makers who were deprived of prioritized values (for example, respect and perceptions of skill) through litigation.

We reached several important conclusions from our mapping and analysis of regional social and decision processes. First, there was little or no opportunity to promote conservation prescriptions through persuasion of regional elected and appointed political officials. Second, litigation entailed significant social costs. Third, there were opportunities – even if limited – for employing social capital to find solutions, based on common ground, to change human practices that were problematic for grizzly bears. In general, opportunities to improve conditions for grizzly bears were severely constrained by the regional social and political context.

We developed seven potential strategic directions or foci based on information about limiting factors and regional social and political conditions. We then scored each strategic direction for relative effectiveness in terms of restoring grizzly bear habitat, and relative tractability in terms of prospects for implementation (see Figure 4.4). We gave highest priority to strategic directions that would yield substantial benefits for grizzly bears and with good prospects for implementation. Second, we sought to identify directions that could be easily implemented and that would yield at least modest gains for bear populations. Based on these considerations, we recommended three strategic elements:

1 using a number of tactics, including education and co-management, developing capacity to change problematic human practices, focusing on sanitation of human facilities, backcountry hunting and camping, and mistaken identity killings – that is, mistaking grizzly bears for black bears (*Ursus americanus*), a legally hunted species;
2 using formal prescriptions as well as principles of co-management, promote road obliteration and closures;
3 using a variety of potential interventions, promoting easements, land purchases and zoning to affect the distribution of humans relative to important grizzly bear habitat.

This example demonstrates a multi-scale effort that considered biological, social and political factors, with the goal of identifying practical strategies that would yield human practices benefiting grizzly bears in areas critical to broad-scale conservation. This example does not contain all of the elements of POCD described above, nor do we necessarily espouse this particular approach for any other situation. Rather, we hope to have provided a tangible example of what facets of a policy-oriented conservation design effort might look like, albeit rooted in somewhat conventional conservation goals that prioritize immediate biophysical considerations. This works, as far as it goes, to frame the issue primarily in terms of biophysical conditions and in ways potentially useful to policy processes. However, this design and others of its type in most cases probably do not provide a reasonable assurance of preserving biodiversity for the indefinite future.

A concluding thought: POCD and governance

Conservation has probably been impeded more by insufficient societal valuation of conservation outcomes and by policy processes biased against conservation-related values than by lack of knowledge about biophysical systems (Brulle, 2000; Carter, 2001; Hayward, 1998). The development of sophisticated CADs and costly monitoring and modelling programmes provides no assurance that the generated information will be used or that conservation goals will be served. Perhaps it is self-evident that governance needs to be an important focus for conservation design, especially as a means of dealing with the long-term uncertainties of complex systems.

Conservation of biodiversity is most likely to be achieved by creating durable systems of governance based on democratic processes that promote common interests and an equitable distribution of values (Brulle, 2000; Brunner, 2002; Clark, 2002). This ideal is *not* represented by current structures that defer to special interests organized around short-term power and wealth values and anthropocentric myths. This is not to say that deference to another set of special interests is the answer. Even if ardent environmentalists were somehow given the authority to impose a drastic conservation agenda, the resulting value deprivations imposed on others would be severe enough to guarantee an eventual loss of authority and, in the longer term, potentially severe damage to conservation interests (Brulle, 2000). It is in an honest reckoning of value indulgences and deprivations, and in a sincere focus on common interests, that values associated with conservation will probably not only be given their due over the long haul, but also be given increasing priority.

If so, then POCD logically focuses on – among other things – crafting sustainable or otherwise 'healthy' systems of governance at all scales, especially if the goal is long-term protection of biodiversity and maintenance of healthy ecosystems. This realization is already implied or expressed outright in many mission statements of conservation organizations. For example, the Y2Y Conservation Initiative includes both cooperation and ecosystem integrity in a mission statement that advocates 'people working together' to create 'an interconnected web of life, capable of supporting all of the natural and human communities that reside within it, for now and for future generations' (www.y2y.net). Whether stated outright or not, the goal is to create or maintain 'healthy' ecosystems *and* 'healthy' human societies. We cannot foresee exactly what will be required to conserve biodiversity in any particular place some distant time in the future. However, the creation and sustenance of equitable policy and decision processes will probably come as close as any measure can to providing for appropriate adaptive responses.

References

Aiken, S. R. (1994) 'Peninsular Malaysia's protected areas' coverage, 1903–92: Creation, rescission, excision and intrusion', *Environmental Conservation*, vol 21, pp49–56

Argyris, C. and Schön, D. A. (1978) *Organizational Learning: A Theory of Action Perspective*, Reading, MA, Addison-Wesley

Barany, M. E., Hammett, A. L., Murphy, B. R. and McCrary, J. K. (2002) 'Resource use and management of selected Nicaraguan protected areas: A case study from the Pacific region', *Natural Areas Journal*, vol 22, pp61–69

Boyce, M. S. and Waller, J. S. (2003) 'Grizzly bears for the Bitterroot: Predicting potential abundance and distribution', *Wildlife Society Bulletin*, vol 31, pp670–683

Brewer, G. D. (1981) 'Where the twain meet: Reconciling science and politics in analysis', *Policy Sciences*, vol 13, pp269–279

Bromley, D. W. and Paavola, J. (eds) (2002) *Economics, Ethics, and Environmental Policy: Contested Choices*, Malden, MA, Blackwell

Brosius, J. P., Lowenhaupt Tsing, A. and Zerner, C. (1998) 'Representing communities: Histories and politics of community-based natural resource management', *Society and Natural Resources*, vol 11, pp157–168

Brulle, R. J. (2000) *Agency, Democracy, and Nature: The US Environmental Movement from a Critical Theory Perspective*, Cambridge, MA, MIT Press

Brunner, R. D. (2002) 'Problems of governance', in Brunner, R. D. Colburn, C. H. Cromley, C. M. Klein, R. A. and Olson, E. A. (eds) *Finding Common Ground: Governance and Natural Resources in the American West*, New Haven, CT, Yale University Press

Carter, N. (2001) *The Politics of the Environment: Ideas, Activism, Policy*, Cambridge, UK, Cambridge University Press

Cawley, R. M. (1993) *Federal Land, Western Anger: The Sagebrush Rebellion and Environmental Politics*, Lawrence, University of Kansas Press

Ceballos, G. and Brown, J. H. (1995) 'Global patterns of mammalian diversity, endemism, and endangerment', *Conservation Biology*, vol 9, pp559–568

Clark, J. N. and McCool, D. C. (1996) *Staking Out the Terrain: Power and Performance Among Natural Resource Agencies*, 2nd edn, Albany, State University Press of New York

Clark, T. W. (1992) 'Practicing natural resource management with a policy orientation', *Environmental Management*, vol 16, pp423–433

Clark, T. W. (1993) 'Creating and using knowledge for species and ecosystem conservation: Science, organizations, and policy', *Perspectives in Biology and Medicine*, vol 36, pp497–525

Clark, T. W. (1997) *Averting Extinction: Reconstructing Endangered Species Recovery*, New Haven, CT, Yale University Press

Clark, T. W. (2000) 'Interdisciplinary problem solving in endangered species conservation: The Yellowstone grizzly bear case', in Reading, R. P. and Miller, B. (eds) *Endangered Animals: A Reference Guide to Conflicting Issues*, Westport, CT, Greenwood Press

Clark, T. W. (2002) *The Policy Process: A Practical Guide for Natural Resource Professionals*, New Haven, CT, Yale University Press

Clark, T. W. and Wallace, R. L. (2002a) 'The professional in endangered species conservation: An introduction to standpoint clarification', *Endangered Species Update*, vol 19, pp101–105

Clark, T. W and Wallace, R. L. (2002b) 'Understanding the human factor in endangered species recovery: an introduction to human social process', *Endangered Species Update*, vol 19, pp87–94

Collar, N. J., Wege, D. C. and Long, A. J. (1997) 'Patterns and causes of endangerment in the New World avifauna', *Ornithological Monographs*, vol 48, pp237–260

Culhane, P. J. (1981) *Public Lands Politics: Interest Group Influence on the Forest Service and the Bureau of Land Management*, Baltimore, MA, Johns Hopkins University Press

Czech, B., Krausman, P. R. and Devers, P. K. (2000) 'Economic associations among causes of species endangerment in the United States', *BioScience*, vol 50, pp593–601

Dasgupta, P. (2001) *Human Well-Being and the Natural Environment*, New York, Oxford University Press

Debinski, D. M. and Holt, R. D. (2000) 'A survey and overview of habitat fragmentation experiments', *Conservation Biology*, vol 14, pp342–355

De Lopez, T. T. (2001) 'Brave new parks: Lessons from internationally funded protected areas in Cambodia', *Natural Areas Journal*, vol 21, pp378–385

Dery, D. (1984) *Problem Definition in Policy Analysis*, Lawrence, University of Kansas Press

Edmunds, S. W. (1987) 'Environmental policy: Bounded rationality applied to unbounded ecological problems', in Mann, D. E. (ed.) *Environmental Policy Formation: The Impacts of Values, Ideology, and Standards*, Lexington, MA, Lexington Books

Ehrlich, P. R. (2002) 'Human natures, nature conservation, and environmental ethics', *BioScience*, vol 52, pp31–43

Formaini, R. (1990) *The Myth of Scientific Public Policy*, New Brunswick, NJ, Transaction Publishers

Fukuyama, F. (2000) 'Social capital', in Harrison, L. E. and Huntington, S. P. (eds) *Culture Matters: How Values Shape Human Progress*, New York, Basic Books

Gagnon Thompson, S. C. and Barton, M. A. (1994) 'Ecocentric and anthropocentric attitudes toward the environment', *Journal of Environmental Psychology*, vol 14, pp149–157

Gareth, M. (1986) 'Organizations as information processing brains', in *Images of Organization*, Thousand Oaks, CA, Sage Publications

Gigerenzer, G. and Goldstein, D. G. (1996) 'Reasoning the fast and frugal way: Models of bounded rationality', *Psychological Review*, vol 103, pp650–670

Gowda, M. V. R. (1999) 'Heuristics, biases, and the regulation of risk', *Policy Sciences*, vol 32, pp59–78

Groves, C. R. (2003) *Drafting a Conservation Blueprint: A Practitioner's Guide to Planning for Biodiversity*, Washington DC, Island Press

Hansen, A. J., Rasker, R., Maxwell, B., Rotella, J. J., Johnson, J.D., Wright Parmenter, A., Langner, U., Cohen, W. B., Lawrence, R. L. and Kraska, M. P. V. (2002) 'Ecological causes and consequences of demographic change in the New West', *BioScience*, vol 52, pp151–162

Hastie, R. and Dawes, R. M. (2001) *Rational Choice in an Uncertain World: The Psychology of Judgment and Decision Making*, Thousand Oaks, CA, Sage Publications

Hawksworth, D. L. (ed.) (1995) *Biodiversity: Measurement and Estimation*, London, Chapman and Hall

Hayward, T. (1998) *Political Theory and Ecological Values*, New York, St Martin's Press

Heinen, J. T. and Low, R. S. (1992) 'Human behavioural ecology and environmental conservation', *Environmental Conservation*, vol 19, pp105–116

Hendee, J. C. and Dawson, C. P. (eds) (2002) *Wilderness Management*, 3rd edn, Golden, CO, Fulcrum Press

Hendee, J. C. and Mattson, D. J. (2002) 'Wildlife in wilderness: A North American and international perspective', in Hendee, J. C. and Dawson, C. P. (eds) *Wilderness Management*, 3rd edn, Golden, CO, Fulcrum Publishing

Homer-Dixon, T. F. (1999) *Environment, Scarcity, and Violence*, Princeton, NJ, Princeton University Press

Inamdar, A., de Jode, H., Lindsey, K. and Cobb, S. (1999) 'Capitalizing on nature: Protected area management', *Science*, vol 283, pp1856–1857

Ingram, H. M. (1973) 'Information channels and environmental decision making', *Natural Resources Journal*, vol 13, pp150–169

Karl, T. R. and Trenberth, K. E. (2003) 'Modern global climate change', *Science*, vol 302, pp1719–1723

Keiter, R. B. and Locke, H. (1996) 'Law and large carnivore conservation in the Rocky Mountains of the U.S. and Canada', *Conservation Biology*, vol 10, pp1003–1012

Kellert, S. R. (1989) 'Perceptions of animals in America', in Hoage, R. J. (ed.) *Perceptions of Animals in American Culture*, Washington DC, Smithsonian Institution Press

Kellert, S. R., Black, M., Reid Rush, C. and Bath, A. J. (1996) 'Human culture and large carnivore conservation in North America', *Conservation Biology*, vol 10, pp977–990

Kerr, J. T. and Currie, D. J. (1995) 'Effects of human activity on global extinction risk', *Conservation Biology*, vol 9, pp1528–1538

Kravtsov, Y. A. (ed.) (1993) *Limits of Predictability*, Berlin, Springer-Verlag

Lachapelle, P. R., McCool, S. F. and Patterson, M. E. (2003) 'Barriers to effective natural resource planning in a "messy" world', *Society and Natural Resources*, vol 16, pp473–490

Lasswell, H. D. (1950) *Politics: Who Gets What, When, How*, New York, Peter Smith

Lasswell, H. D. (1971) *A Pre-View of Policy Sciences*, New York, American Elsevier

Lasswell, H. D and McDougal, M. (eds) (1992) *Jurisprudence for a Free Society: Studies in Law, Science and Policy*, New Haven, CT, New Haven Press

Lee, K. N. (1993) *Compass and Gyroscope: Integrating Science and Politics for the Environment*, Washington DC, Island Press

Lemons, J. (1996) *Scientific Uncertainty and Environmental Problem Solving*, Cambridge, MA, Blackwell Science

Lichatowich, J. (1999) *Salmon Without Rivers: A History of the Pacific Salmon Crisis*, Washington DC, Island Press

Ludwig, D., Hilborn, R. and Walters, C. (1993) 'Uncertainty, resource exploitation, and conservation: Lessons form history', *Science*, vol 260, pp17–36

Lusigi, W. J. (1994) 'Socio-economic and ecological prospects for multiple use of protected areas in Africa', *Biodiversity and Conservation*, vol 3, pp449–458

Machlis, G. E., Force, J. E. and Burch, W. R. Jr. (1997) 'The human ecosystem part I: The human ecosystems as an organizing concept in ecosystem management', *Society and Natural Resources*, vol 10, pp347–367

Manning, R., Valliere, W. and Minteer, B. (1999) 'Values, ethics, and attitudes toward National Forest management: An empirical study', *Society and Natural Resources*, vol 12, pp421–436

Markman, A. B. and Gentner, D. (2001) 'Thinking', *Annual Review of Psychology*, vol 52, pp223–247

Mattson, D. J. and Merrill, T. (2002) 'Extirpations of grizzly bears in the contiguous United States, 1850–2000', *Conservation Biology*, vol 16, pp1123–1136

Mattson, D. J. and Merrill, T. (2004) 'A model-based appraisal of habitat conditions for grizzly bears in the Cabinet-Yaak region of Montana and Idaho', *Ursus*, vol 15, pp78–91

McDougal, M. S., Lasswell, H. D. and Reisman, W. M. (1981) 'The intelligence function and world public order', in McDougal, M. S. and Reisman, W. M. (eds)

International Law Essays, Mineola, NY, Foundation Press

McKinney, M. L. (2001) 'Role of human population size in raising bird and mammal threat among nations', *Animal Conservation*, vol 4, pp45–57

Mellers, B. A., Schwartz, A. and Cooke, A. D. J. (1998) 'Judgment and decision making', *Annual Review of Psychology*, vol 49, pp447–477

Merrill, T. (2004) *Conservation Area Design for Grizzly Bears of the Yellowstone to Yukon Ecoregion*, Canmore, AB, Yellowstone to Yukon Initiative

Merrill, T. and Mattson, D. J. (2003) 'The extent and location of habitat biophysically suitable for grizzly bears in the Yellowstone region', *Ursus*, vol 14, pp171–187

Merrill, T., Mattson, D. J., Wright, R. G. and Quigley, H. B. (1999) 'Defining landscapes suitable for restoration of grizzly bears *Ursus arctos* in Idaho', *Biological Conservation*, vol 87, pp231–248

Miles, E. L. (1998) 'Personal reflections on an unfinished journey through global environmental problems of long timescale', *Policy Sciences*, vol 31, pp1–33

Miller, A. (1999) *Environmental Problem Solving: Psychosocial Barriers to Adaptive Management*, New York, Springer

Miller, B., Reading, R., Conway, C., Jackson, J. A., Hutchins, M., Snyder, N., Forrest, S., Frazier, J. and Derrickson, S. (1994) 'A model for improving endangered species recovery programs', *Environmental Management*, vol 18, pp637–645

Miller, B., Reading, R. and Forrest, S. (1996) *Prairie Night: Black-footed Ferrets and the Recovery of Endangered Species*, Washington DC, Smithsonian Institution Press

Noss, R. F. (2003) 'A checklist for wildlands network designs', *Conservation Biology*, vol 17, pp1270–1275

Noss, R. F. and Cooperrider, A. Y. (1994) *Saving Nature's Legacy: Protecting and Restoring Biodiversity*, Washington DC, Island Press

Noss, R. F., O'Connell, M. A. and Murphy, D. D. (1997) *The Science of Conservation Planning: Habitat Conservation under the Endangered Species Act*, Washington DC, Island Press

Oskamp, S. (1995) 'Applying social psychology to avoid ecological disaster', *The Journal of Social Issues*, vol 51, pp217–239

Parker, V. (1995) 'Natural resources management by litigation', in Knight R. L. and Bates S. F. (eds) *A New Century for Natural Resources Management*, Washington, DC, Island Press

Phillips, K. P. (2002) *Wealth and Democracy: A Political History of the American Rich*, New York, Broadway Books

Portney, R. R. and Weyant, J. P. (1999) *Discounting and Intergenerational Equity*, Washington DC, Resources for the Future

Puttaswamaiah, K. (ed.) (2002) *Cost–Benefit Analysis: Environmental and Ecological Perspectives*, New Brunswick, NJ, Transaction Publishers

Ravetz, J. R. (1996) *Scientific Knowledge and its Social Problems*, New Brunswick, NJ, Transaction Publishers

Reading, R. P., Clark, T. W. and Kellert, S. R. (1994) 'Attitudes and knowledge of people living in the Greater Yellowstone Ecosystem', *Society and Natural Resources*, vol 7, pp349–365

Reading, R. P and Kellert, S. R. (1993) 'Attitudes toward a proposed reintroduction of black-footed ferrets *(Mustela nigripes)*', *Conservation Biology*, vol 7, pp569–580

Redford, K. H., Coppolillo, P., Sanderson, E. W., DaFonseca, G. A. B., Dinerstein, E., Groves, C., Mace, G., Magginnis, S., Mittermeier, R. A., Noss, R., Olson, D., Robinson, J. G., Vedder, A. and Wright, M. (2003) 'Mapping the conservation landscape', *Conservation Biology*, vol 17, pp116–131

Reiger, J. F. (2001) *American Sportsmen and the Origins of Conservation*, 3rd edn, Corvallis, Oregon State University Press

Reisner, M. (1993) *Cadillac Desert: The American West and its Disappearing Water*, New York, Penguin Books

Sabatier, P. (1978) 'The acquisition and utilization of technical information by administrative agencies', *Administrative Science Quarterly*, vol 23, pp396–417

Salafsky, N., Margoluis, R., Redford, K. H. and Robinson, J. G. (2002) 'Improving the practice of conservation: A conceptual framework and research agenda for conservation science', *Conservation Biology*, vol 16, pp1469–1479

Saunders, D. A., Hobbs, R. J. and Margules, C. R. (1991) 'Biological consequences of ecosystem fragmentation: A review', *Conservation Biology*, vol 5, pp18–32

Shafir, E. and LeBoeuf, R. A. (2002) 'Rationality', *Annual Review of Psychology*, vol 53, pp491–517

Shrader-Frechette, K. S. (1991) *Risk and Rationality*, Berkeley, University of California Press

Shrader-Frechette, K. S. and McCoy, E. D. (1993) *Method in Ecology: Strategies for Conservation*, Cambridge, UK, Cambridge University Press

Singleton, S. (2000) 'Co-operation or capture? The paradox of co-management and community participation in natural resource management and environmental policy-making', *Environmental Politics*, vol 9, pp1–21

Smutny, G. and Takahashi, L. (1999) 'Economic change and environmental conflict in the western mountain states of the USA', *Environment and Planning A*, vol 31, pp979–995

Solimano, A. (ed.) (1998) *Social Inequality: Values, Growth, and the State*, Ann Arbor, University of Michigan Press

Somit, A. and Peterson, S. A. (1997) *Darwinism, Dominance, and Democracy: The Biological Bases of Authoritarianism*, Westport, CT, Praeger

Steelman, T. A. and Ascher, W. (1997) 'Public involvement methods in natural resource policy making: Advantages, disadvantages and trade-offs', *Policy Sciences*, vol 30, pp71–90

Taylor, J. E. III. (1999) *Making Salmon: An Environmental History of the Northwest Fisheries Crisis*, Seattle, University of Washington Press

Thiele, L. P. (1999) *Environmentalism for a New Millennium: The Challenge of Coevolution*, New York, Oxford University Press

Wagner, F. H., Foresta, R., Gill, R. B., McCullough, D. R., Pelton, M. R., Porter, W. F. and Salwasser, H. (1995) *Wildlife Policies in the US National Parks*, Washington DC, Island Press

Weber, E. P. (1998) *Pluralism by the Rules: Conflict and Cooperation in Environmental Regulation*, Washington DC, Georgetown University Press

Wilkinson, T. (1998) *Science Under Siege: The Politician's War on Nature and Truth*, Boulder, CO, Johnson Books

Wilshusen, P. R., Brechin, S. R., Fortwangler, C. L. and West, P. C. (2002) 'Reinventing a square wheel: A critique of a resurgent "protection paradigm" in international biodiversity conservation', *Society and Natural Resources*, vol 15, pp17–40

Wilson, J. Q. (1989) *Bureaucracy: What Government Agencies Do and Why They Do it*, New York, Basic Books

Wondolleck, J. M. and Yaffee, S. L. (2000) *Making Collaboration Work: Lessons from Innovation in Natural Resource Management*, Washington DC, Island Press

Wright, R. G. and Mattson, D. J. (1996) 'The origin and purpose of national parks and protected areas', in Wright R. G. (ed.) *National Parks and Protected Areas: Their Role in Environmental Protection*, Cambridge, MA, Blackwell Science

Yaffee, S. L. (1994) *The Wisdom of the Spotted Owl: Policy Lessons from a New Century*, Washington DC, Island Press

Appendix: An outline for some facets of policy-oriented conservation design

The following is a step-wise more-or-less sequential approach to conservation design labelled by phases of the policy process internal to the organization motivating the design. Intelligence phases are ideally conducted by scientists and other disciplinary specialists. Prescriptive phases are overtly goal-oriented and value-driven.

The outline

Figuring out what is wanted and what the focus should be

Goals

Prescription	1.	Articulate specific conservation goals (contextually explicit goals).

Translation of goals into functional elements

Prescription	2.	Identify specific biophysical elements that are linked by the organization with the attainment of goals (organizationally relevant biophysical elements).
Intelligence	3.	Clarify the norms and values embodied in these elements.
Intelligence	4.	Identify specific biophysical elements or related metrics that are linked by scientific theory with the attainment of stated goals.
Appraisal	5.	Clarify the overlap and synonymy between 2 and 4 and, if warranted, revisit 1 and 2 and revise in light of 3 and 4.

Figuring out where to focus and what to do there

Biophysical elements: Modelling and prioritizing elements of the biophysical system

Intelligence	6.	Identify *key proximal stressors* or limiting factors for each biophysical element that is either or both organizationally and scientifically relevant.
Intelligence	7.	Identify *specific human-related activities and landscape features* that have a major governing effect on proximal stressors.

Intelligence	8.	Determine the *broader cultural and societal significance* and value of each biophysical element; to whom and where.
Intelligence	9.	Describe similarities among biophysical elements in terms of similarities of sensitivities to key stressors and, more importantly, related human activities and landscape features.
Intelligence	10.	Rank order the elements of each group by (a) amenability to modelling, (b) scientific meaning, (c) meaning to the conservation organization, and (d) broader cultural and societal significance.
Intelligence	11.	Focusing on top-ranked elements, develop spatially and temporally explicit models of linkages between element states and stressors and/or human-related factors.
Intelligence	12.	Identify human activities and human-related landscape features that engender, cumulatively over all key biophysical elements, the greatest stress (high impact stressors).

Biophysical elements: Appraising the biophysical system

Intelligence	13.	Map the current distribution of high impact stressors; determine trends and projections for each.
Intelligence	14.	Map current conditions for each key biophysical element, by modelling or direct observation; determine trends and projections for each.
Prescription	15.	Establish functional goals and thresholds for each stressor and biophysical element.
Intelligence	16.	Determine the location and extent of acceptable conditions for each stressor and biophysical element; articulate in functional terms what elements of these areas need to be preserved to maintain them in an acceptable state.
Appraisal	17.	Determine whether landscape-level goals have been met under current conditions for each stressor/ biophysical element.
Intelligence	18.	If not, identify substandard areas key to achieving broad landscape-level objectives and amenable to restoration.

Intelligence	19.	Given future projections of key stressors, identify areas at greatest risk of deterioration and the reasons why, for each stressor and biophysical element.

Biophysical elements: Identifying and characterizing preservation planning areas

Intelligence	20.	Determine the location and extent of overlap among currently acceptable conditions for key stressors and key biophysical elements.
Intelligence	21.	Identify areas of high frequency of overlap of acceptable conditions.
Intelligence	22.	Identify the subset of these areas at risk owing to projected increases in stressors.
Intelligence	23.	Clarify the amalgam of attributes (landscape conditions) that determine the current value of these areas and that require preservation.
Prescription	24.	Delineate preservation planning areas and prescribe elements to be preserved in each.

Biophysical elements: Identifying and characterizing restoration planning areas

Intelligence	25.	Determine the location and extent of overlap among high priority restoration areas for each stressor and biophysical element.
Intelligence	26.	Identify areas of high frequency of overlap among high priority restoration areas for each stressor and biophysical element.
Intelligence	27.	Clarify/determine the amalgam of restoration actions that need to be undertaken to achieve goals for all stressors and biophysical elements in the areas of greatest overlap.
Prescription	28.	Delineate restoration planning areas and prescribe elements to be restored in each.

Figuring out how to get change where it's wanted

Mapping ownership

Intelligence	1.	For each preservation and restoration planning area, map ownerships.

Invocation and implementation of existing policies

Intelligence   2.   For each mapped and prioritized ownership, compile and analyse policies, formal and otherwise, governing management of stressors and biophysical elements.

Intelligence   3.   Map the social process affecting invocation and implementation of policies or normal practices for each relevant land ownership.
  A. Identify participants and create a profile for each that describes identities, strategies, expectations, demands and so on.
  B. Describe relations among participants including the strength and nature of effects and the existence and nature of alliances.
  C. Conduct this analysis at local, regional, national and international scales, as relevant, including cross-scale linkages.

Intelligence   4.   Determine who needs to make what decisions, when and where under current policies to elicit desired changes in biophysical conditions – including options on several pathways.

Prescription   5.   Determine a strategy; who does what, when and where to make things happen.

Implementation   6.   Implement the strategy.

Intelligence   7.   Concurrently, estimate the adequacy up front of existing policies, formal or otherwise, to achieve goals for preservation and restoration areas.

Prescription of new policies, termination of existing ones

Prescription   8.   Articulate new policies needed to achieve goals and identify existing detrimental policies that need to be terminated.

Intelligence   9.   Identify governmental entities or other organizations that are best suited to prescribing or terminating targeted policies.

Intelligence   10.   Map the social process affecting prescription and termination of policies (see above).

Intelligence   11.   Determine who needs to make what decisions, when and where, to prescribe or terminate targeted policies.

Prescription   12.   Determine a strategy; who does what, when and where to make things happen.

Implementation 13.   Implement the strategy.

Learning

1.   Articulate standards and identify related metrics for appraising progress and achieving success.

2.   Appraise progress and the success of strategies.

5

Introducing Innovations into Watershed Management

Sandra Brown and Hans Schreier

Introduction

In March 2005, the UN proclaimed the 'Water for Life' decade in an attempt to draw attention to the rapidly emerging water problems around the world. Providing sufficient and safe water to all users while maintaining sufficient water for aquatic biota and ecosystems is probably one of the greatest human challenges we face in this century. Since the major freshwater sources originate in mountains (Viviroli et al, 2003), it is essential that we sustain and protect the water resources in mountains. This can only be achieved with the introduction of innovative approaches to watershed management. In view of increasing land-use intensification and climatic variability, it is clear that traditional management approaches are no longer adequate.

The aim of this chapter is, first, to review traditional approaches to mountain watershed management and, second, to suggest how innovative management practices can be used to provide some protection against the anticipated increases in water quality deteriorations associated with land-use changes and increased variability in supply associated with climatic change. Experiences from several mountain watersheds are used to document effective ways of improving the availability and quality of the water resources. Because water resources are dynamic, an interdisciplinary team approach is essential if we hope to be able to solve the wide range of problems associated with water management.

Problems with traditional approaches to watershed management

Managing water has primarily been the task of engineers, and they have been very efficient and effective in developing suitable infrastructure that retains and stores water, delivers it for irrigation and domestic use, drains and conveys stormwater, and in constructing water treatment facilities. This is usually referred to as the 'hard path' of water management (Wolff and Gleick, 2002). While these systems have been effective, they have not been well integrated in a watershed context and were usually constructed in a piecemeal fashion without sufficient consideration of cumulative effects. Historic hydrometric data usually form the basis for the designs of structures and, in order to reduce the effect of uncertainty and the risk of failure, the systems were designed with an increased safety factor.

There is now enough evidence to suggest that climatic extremes are increasing due to global warming (Labat et al, 2004). What makes this even more challenging is the fact that land-use intensification in most watersheds has also altered the hydrological regime, particularly rainfall runoff processes. The creation of impervious surfaces for transportation and urbanization, and the compaction of soils associated with forestry, agriculture and recreation has resulted in significant decreases in water infiltration, recharge and water storage in soils and groundwater aquifers. This leads to higher rates of surface runoff during storm events and more droughty conditions during the dry season. At the same time, impervious surfaces have accelerated the transport of pollutants through surface flow into streams and groundwater, a phenomenon that is particularly evident in urban environments. The combination of increased climatic variability and surface alteration has not been addressed sufficiently using the traditional engineering approaches; these two factors are now emerging as key challenges facing water managers.

Making the switch from tradition to innovation

It is now well established that no single intervention will protect water resources at all times, for all uses, and against all pollutants. This is evident for drinking water, where the focus has shifted from treatment to source control and source-to-tap protection. The protection of human and ecosystem health in water resources management needs to consider a multi-barrier watershed approach. Some of the traditional land-use practices that impact on the quality and quantity of water are listed in Table 5.1. To minimize these impacts requires innovative measures that often represent a complete reversal from the traditional approach or a major shift in emphasis.

All of these changes are needed as water demands increase in all sectors of the economy, and water pollution from non-point sources becomes more widespread (Leaf and Chatterjee, 1999; Ritter and Shirmohammadi, 2001). We now have two new concerns: how much water should we leave in streams to maintain aquatic biota and environmental services (Postel and Richter, 2003); and what is our response to increased climatic variability due to global warming? Since we have relatively little experience in dealing with climate change and water allocations for environmental services, it is essential to take an adaptive management approach when introducing the proposed innovations. We need

Table 5.1 *A comparison between traditional and innovative approaches to watershed management*

	Traditional Approach	Innovative Approach
Changes needed in practices	Create impervious surfaces (pavement) and compacted soils	Minimize impervious surfaces and soil compaction, maintain infiltration capacity
	Minimize riparian buffer zones	Maximize riparian buffer zones
	Drain wetlands	Create wetlands
	Manage blue water	Manage green water
	Engineered river channels	Return channels to natural channels
Changes needed in focus	Point source pollution	Non-point source pollution
	Expand water supplies	Reduce water demand (conservation)
	Treat water	Source control
	Sector-based management	Integrated watershed management
	Water primarily for human use	Water allocations for environmental services
	Government management	Community-based management

to include and modify some of the 'hard path' practices and put more emphasis on the 'soft path' approach. 'Soft path' practices focus on education and behaviour change to improve water/watershed management, and thus complement traditional engineering options to use water more wisely and efficiently.

Combating soil compaction and impervious surfaces

Most land-use activities compact soils and, as a result, soils are losing part of their natural water storage capacity. With increased climatic extremes, this leads to more surface runoff, soil erosion and losses of the most productive part of the soil profile. If we make soil surfaces impervious by paving a large proportion of the landscape, we change the hydrological response by increasing peak flows in rivers after storm events, so the lag time between rainfall and runoff will be more rapid and the base flow will generally be reduced. At the same time, pollutants that accumulate on impervious surfaces during dry spells will be conveyed more rapidly into streams during storm events, leading to widespread contamination and toxicity. This is usually referred to as the first flush event, which places tremendous stress on the aquatic environments (Lee et al, 2002). Numerous studies (Arnold and Gibbons, 1996; Booth et al, 2002; Horner et al, 1997) have shown that aquatic biodiversity is severely impaired as the impervious area in a watershed increases.

Horner et al (1997) used the Integrated Benthic Index (IBI) and other biodiversity indicators to show that 10–12 per cent imperviousness represents the threshold beyond which aquatic biodiversity decreases. Typical urban watersheds have between 40 and 60 per cent imperviousness and, once established, it is difficult to reverse these characteristics. The cause of the decline in fish and biodiversity is not the amount of imperviousness *per se*, but a combination of a more flashy hydrological regime and the pulses of contamination that are introduced during storm events as a result of increased surface runoff from paved surfaces. As a result, imperviousness in watersheds has now become a key environmental indicator (Arnold and Gibbons, 1996; Schuler, 1994). If we hope to maintain a healthy aquatic ecosystem, all new development should be designed in such a way as not to exceed 10–12 per cent of impervious surfaces.

Solutions to the imperviousness and compaction problem are well known, but the challenge is how to translate this knowledge into action. Soil compaction in agriculture can be minimized by maintaining good organic matter content in the soils, preventing soil surface crusting, minimizing ploughing (low till), and maintaining cover crops during critical times of the year. This is particularly important in mountain headwaters where soil moisture holding capacity influences downstream base flow. In the urban environment, we need to promote pervious pavement, such as parking lots with various sand and grass infill between honey-cone grids and cobblestones. This not only decreases the surface runoff but means that the oil, hydrocarbons and metals that accumulate in parking spaces will infiltrate into the soil where chemical and biological processes have a chance to modify, adsorb and decompose contaminants. Additional options in the urban environment include reducing residential street width, redirecting rooftop runoff to yards, and grass swales instead of piped roadside drainage ditches.

Enlarged riparian buffer zones along streams

It has been well established (Castelle et al, 1994) that wide vegetated buffer zones are an effective mechanism to prevent pollutants, sediments and excess nutrients from entering streams. To be effective, such zones should be continuous and at least 30–60m wide. The services that they can provide include noise protection; shading to maintain low water temperatures; expanded habitat for organisms and wildlife; supply of large woody debris that enhances food supply and provides stream channel diversity; retention of pollutants within the terrestrial zone for detention; plant take-up; and decomposition. Buffer zones are also an effective barrier for flood control. A combination of grasses, shrubs and trees has proven to be most effective; the width of the buffer can be adjusted depending on the channel morphology and the surficial material composition of the buffer corridor. In headwater areas of mountain streams, sufficient and continuous riparian vegetation will aid in providing improved water quality for downstream users.

Creating wetlands

Humankind has a long history of draining wetlands although they are among the most productive ecosystems on Earth. Wetlands are among the most effective water storage systems, they act as very efficient filters and they are capable of absorbing and bio-remediating pollutants and taking up excess nutrients (Sakadevan and Bavor, 1999). Wetlands function as flow regulators and are even capable of reducing microbial contamination. Draining wetlands has been popular in the past to ease construction and transport problems and to transfer the land into crop production that usually requires well-drained conditions, with the organic matter being used as a valuable soil amendment. In mountains, draining wetlands for activities such as cattle ranching is problematic as the entire downstream hydrologic regime may be affected.

Recognizing the value of all the services that wetlands provide, they are now being created not only in the rural environment, to enhance habitat and act as filters for non-point sources of pollution from agriculture, but also in the urban environment (Birch et al, 2004). Wetlands are proving to be very effective in urban stormwater management where they function both as stormwater storage devices and as retention basins for urban contaminants. Mallin et al (2002) and Brydon (2004) have shown that metal contamination in streams can be reduced by 30–50 per cent if the urban stormwater passes through a stormwater detention system before it enters the stream. These systems are most effective if they contain wetland vegetation. It is essential that we maintain wide buffer zones, which provides us the space to create and conserve such systems within the stream buffer zone.

Managing green water in addition to blue water

Blue water is the component of the rainfall that becomes runoff, lake and streamflow, and groundwater (Rockstrom and Gordon, 2001). This is the part of the hydrological cycle that is the main source of water for human use. Green water is the component of rainwater that is intercepted by the vegetation or enters the soil, is taken up by biomass and evapo-transpired back into the atmosphere. There is roughly the same amount of water in the vegetation/soil component of the hydrological cycle as is in the blue cycle, but we have not done much to manage the water efficiently in the green cycle. If we hope to produce more food, fibre and biomass, we need to start paying much more attention to improving the efficiency of the 'conversion' of rainwater into biomass.

Much of this can be accomplished by promoting plants that are efficient water converters and are grown under climatic conditions suited for these specific plants (Rockstrom, 2003). C4 plants appear to be more effective than C3 plants, and we need to develop a much better knowledge base on the water needs and conversion rates for different tree species.*

---

* C3 and C4 are metabolic pathways associated with the photosynthesis fixation of carbon dioxide. C4 plants utilize the C4 photosynthetic carbon assimilation cycle, which decreases water vapour loss and enables C4 plants to be more water efficient than C3 plants, which utilize the photosynthetic carbon reduction cycle.

Converting dammed and channelled streams back to natural streams

To provide storage for irrigation and hydropower development, to 'tame' streams, regulate flows and protect lands from floods, it has been a common practice to channelize streams and build storage reservoirs. However, in the long term, due to problems with sedimentation and increased climatic variability, many of these storage and channelized systems do not easily perform their intended functions under these new conditions (Khagram, 2004). The impacts of dams on upstream and downstream flow and stream morphology are summarized in Table 5.2. As pointed out by Rosenberg et al (2000), these structural modifications of streams have had a global impact on hydrology.

Table 5.2 *Impacts of dams, reservoirs and constructed channels on hydrology, water quality and stream habitat*

Impacts of dams and reservoirs	
Upstream impacts	**Downstream impacts**
Displacement of people, loss of valuable land	Regulation of streamflow regime
Reservoir shoreline issues (exposure, sand)	Pulsing flow regime
Sediment accumulation	Modified sediment transport regime
Progressive channel degradation	Changes in aquatic biota
Changes in microclimatic conditions	Changes in water temperatures and quality
Inhibits fish migration	Inhibits fish migration, alternative species
Mercury and methane production in reservoirs	Risk of dam failure

Impacts of constructed Channels	
Impacts on flow regime	**Impacts on habitat**
Increased flow velocity	Loss of channel diversity and fish habitat
Increased flood risk during high flow	Less conducive environment for fish
Alteration of sediment transport	Reduced environmental services
Scouring upstream	
Increased sediment deposition downstream	

Many projects are now under way to return engineered stream systems to more natural systems that can expand and contract more effectively than streams that were converted into constrained channels. In some regions, older dams are being demolished because they are no longer viable due to sedimentation, and to improve fish passage and environmental services. Channel conversion back into natural streams is often difficult because of considerable development after channel construction that cannot be relocated due to the high cost. However, to improve environmental services, such conversions are essential; any future canalization of streams should be minimized wherever possible.

Shifting attention from point sources of pollution to non-point sources

Non-point sources (NPSs) of pollution from agriculture and urbanization now far exceed the contribution from point sources (Leaf and Chatterjee, 1999; Ritter and Shirmohammadi, 2001). This is partly due to the emphasis of regulation and treatment requirements on reducing industrial inputs into streams and groundwater. The NPS pollution problem is far more difficult to address because of the complexity created by many disperse small sources and the difficulties of determining each contribution to the cumulative load. Simple regulations do not suffice because each individual contribution is not in violation of established norms. The only way to address the NPS problem is through source reduction, best management practices (BMPs) and to have a multi-barrier watershed protection system that includes all of the above mentioned innovations.

Moving from expanding supplies to controlling demand

Mountains are the 'water towers' for downstream populations, providing more than 60 per cent of freshwater resources from less than 35 per cent of the land base (Viviroli et al, 2003). In many parts of the world, however, there are few unexplored new freshwater supplies and, given global warming, relying on glacial melt is becoming increasingly risky. This suggests that a better option is to focus on water demand management. In both agricultural and urban environments, water-use efficiency has not been given much attention. Gleick et al (2004a, 2004b) suggest that most of the current and expected water shortages in southern California could be avoided if water conservation were given priority. The greatest and most inefficient water sector is agriculture, which uses 70 per cent of the fresh water, and the dominant irrigation technique is flood irrigation. Efficient drip and sprinkler use can reduce water needs by 50–70 per cent (Oweis et al, 1999; von Westarp et al, 2004). Similarly, urban water users, particularly in North America, are prolific water consumers because water as a common pool resource is highly undervalued. It has been estimated that demand can be curtailed by at least 30–50 per cent with the introduction of low water-use facilities (for example, toilets, shower heads, reuse).

Reducing treatment cost by moving to source control

Treating water is becoming an increasingly expensive proposition. We have well-established technologies to deal with sediment, organic matter and biological oxygen demand. However, the removal of excess nutrients, complex organic contaminants, metals and pharmaceuticals is technically challenging and very expensive. It is unlikely that we will be able to remove all pollutants in wastewater treatment systems now or in the future. In order to maintain aquatic and ecosystem health, a much more effective way to reduce pollutant loads into streams is by source control. Reducing inputs at the source can be accomplished in agriculture by determining nutrient budgets on an annual basis. This forces land users to balance inputs with crop needs. Similarly, sediment budgets can be determined, which helps in the identification of sediment sources. By combining source control with BMPs (for example, manure storage, fertilizer timing, cover crops), impacts can be effectively reduced.

Achieving equity between human use and ecosystem health needs

Until recently, water allocations were made primarily for human use; relatively little attention was given to allocating water for maintaining environmental services. However, with awareness of the importance of the services that nature provides, it is now apparent that we need to re-evaluate water allocations and maintain sufficient water flow and a water quality regime in rivers so that they are capable of maintaining fish and aquatic biota throughout the year. As shown by Postel and Richter (2003), this is not an easy task, and the methods that have been developed so far are somewhat simplistic. When comparing the biological attributes with hydrology, the inevitable questions are: what is the minimum requirement for keystone species, and how do we integrate these requirements to give adequate representations to all organisms in the food chain? Tennant's (1975) method suggests that 60–100 per cent of the average flow needs to be protected in order to sustain optimum biological conditions. This is not easily achieved during low-flow summer periods. This idea has evolved into the determination of aquatic base flow standards for many rivers in the US, with a focus on minimum flow in August. Water depth and flow are usually the two most important variables that are considered and, using the 'physical habitat simulation model' (PHABSIM), a biologist can determine the necessary flow levels to create the greatest amount of habitat for any given indicator species.

Ecological services are provided during low base flow and high peak flow, and this means that a pulsing or fluctuating regime needs to be in place to provide the range of services that are necessary for creating a well-functioning ecosystem. Peak flow is responsible for modifying channels, creating new habitat (pools and riffles), flushing out pollutants and sediments, providing new nutrient sources and assisting migration of organisms; while low flow provides opportunities for spawning, and the alteration of temperature, moisture and nutrient regimes that favour reproduction and productivity. By maintaining sufficient minimal flow, we can control water quality, temperatures, water tables and dissolved oxygen levels.

Achieving a flow regime that mimics natural systems is quite contrary to human demands on stream systems. The human approach is to minimize variability, taming rivers by regulating flow using hydro-dams and reservoirs, with the main focus being flood protection and energy generation. Drought protection has not been given sufficient attention, but is probably far more important than floods because its impacts are much greater and prolonged. The new paradigm is to harvest most of the water used for human activity during the wet months and minimize river water use during the dry period, but at the same time to maintain a regime that mimics the natural flow. An alternative approach, as proposed by Postel and Richter (2003), is to set specific ecological goals and then use adaptive management approaches to try to achieve these goals.

From sector-based management
to integrated watershed management

Water management by government agencies is largely sectoral. Government professionals involved in hydropower development, irrigation, drinking water use, fish management and flood protection have given little or insufficient attention to other water uses or functions. This has led to many conflicts that could not be resolved in a satisfactory manner, given the rigid division between the different government departments dealing with water. As the demand for water and climatic variability increases, the various sectors can no longer work in isolation; an integrated watershed approach is required, with water issues being discussed in a multi-stakeholder forum. Governments are often constrained in collaborative efforts due to the large number of agencies that are responsible for certain aspects of water management and their jurisdictional, regulatory and financial constraints. Agencies that monitor water often have little or no say in its allocation, its conservation, health aspects or enforcement. Data sharing has never been easy between agencies. These functions need to be integrated if we hope to arrive at a water management policy that satisfies most users and stakeholders, and minimizes conflicts.

The rapid integration of agencies and the public is needed to create watershed-wide management practices that consider all users' needs, environmental needs and the importance of mountains. This is now starting to come to fruition in certain watersheds where collaboration between government agencies and community groups is taking place in the form of watershed councils. Examples include: the Brunette Basin Task Group in Canada where a multi-stakeholder/multi-jurisdictional forum has been used to develop and implement an integrated stormwater management plan (WLAP, 2002) and a basin watershed plan (GVRD, 2001); the Lingmutey Chhu watershed in Bhutan, which served as a pilot watershed for community-based natural resource management and implementation of national policy where river basins form the national water management system (RNRRC-Bajo, 2002; Saciwaters, 2005); and the Columbia Basin Trust in Canada that supports local efforts to achieve economic, environmental and social benefits for the region, facilitates networking and acts an intermediary between government and local stakeholders (CBT, 2005a, 2005b). However, the success of these initiatives is largely dependent on individuals who are enlightened and have strong leadership qualities. These multi-stakeholder processes are complex and time-consuming but, with the help of innovative leaders and an emphasis on public education, significant progress is being made in many parts of the world.

As watersheds are the optimum landscape units to make successful linkages between land use, hydrology and water quality, an integrated watershed approach is essential. For example, nutrient budgets can be calculated for individual farms so that linkages between land use, nutrient surplus applications and water quality deterioration can be determined at the watershed scale (Berka et al, 2000). Similarly, we can establish water and sediment balances in a dynamic way to provide a scientific basis for allocating and initiating conservation measures for erosion control and sediment transport within a watershed. Many governments are now recognizing the benefit of using watersheds as a planning unit – including those of South Africa (Anderson, 2002), Bhutan (Saciwaters, 2005), New Zealand (Landcare, 2005) and Colombia (República de Colombia, 2004) – but time is required to assess the effectiveness of these relatively new national initiatives. At the local/municipal level, more progress is being made on the ground by innovative leaders. In the township of Langley in British Columbia (BC), Canada, the environmental coordinator of the municipality has teamed with a local stewardship group, the Langley Environmental Partners Society, to develop and implement a water resources management strategy and a water-wise programme (TOL, 2005a, 2005b). In Greater Vancouver and the Lower Fraser Valley of BC, constructed wetlands and detention ponds are being used by numerous municipalities to filter contaminants from subdivisions and roads (Brydon et al, 2006), and research results are being shared locally and internationally (for example, Yorito, Honduras). The key, however, is linking local initiatives into comprehensive action and national policy; a difficult and often unachieved task.

Integrating innovative approaches into mountain headwater systems

The control of land-use activities to minimize impacts on hydrology and water quality has to start in mountain headwater systems. If we cannot control water resources at or near the source, most downstream users and aquatic biota will be adversely affected. No single approach will assure that we can sustain all uses of water and functions that water provides. All human activities have greater or lesser impacts on water, and we strive towards consistency in supply and quality. However, nature operates in a more stochastic manner and we have to learn to find a middle ground between these two opposing trends. Given the expected increases in climatic variability, a promising option is to establish a multi-barrier approach that is focused on minimizing land-use and stream alterations that affect hydrology, initiating conservation measures that reduce water demands, and putting enough protective measures into watersheds that allow for multiple ways to protect streamflow and water quality. To accomplish all of this requires new management structures and innovative approaches, as discussed above, combining hard and soft path strategies.

Each watershed is different and requires a different set of management approaches. The innovative methods proposed in this paper fall into three categories: preventative protection measures, rehabilitation methods and conservation measures. All three will be needed and must be in place in all watersheds, upstream and downstream, if we hope to provide sufficient water of good quality to a world's population that is expected to peak at nine billion people by 2050.

References

Anderson, A. (2002) *An Empowered Voice? An Assessment of the Participatory Process Conducted to Draft a Proposal for the Establishment of the Inkomati Catchment Management Agency Mpmalanga, South Africa*, MSc thesis, Resource Management and Environmental Studies, Vancouver, University of British Columbia

Arnold, C. L. and Gibbons, C. J. (1996) 'Impervious surface coverage: The emergence of a key environmental indicator', *Journal of American Planning Association*, vol 62, pp243–258

Berka, C., Schreier, H. and Hall, K. (2000) 'Linking water quality with agricultural intensification in a rural watershed', *Water, Air and Soil Pollution*, vol 127, pp389–401

Birch, G. F., Matthai, C., Fazeli, M. S. and Shu, J. Y. (2004) 'Efficiency of a constructed wetland in removing contaminants from stormwater', *Wetlands*, vol 24, pp459–466

Booth, D. B., Hartley, D. and Jackson, R. (2002) 'Forest cover, impervious surface area and mitigation of stormwater impacts', *Journal of American Water Resource Association*, vol 38, pp835–845

Brydon, J. (2004) *The Effectiveness of Stormwater Ponds in Contaminant Removal from Urban Stormwater Runoff in the Lower Fraser Valley, BC*, MSc thesis, Resource Management and Environmental Studies, Vancouver, University of British Columbia

Brydon, J., Roa, M. C., Brown, S. J. and Schreier, H. (2006) 'Integrating wetlands into watershed management: Effectiveness of constructed wetlands to reduce impacts from urban stormwater', in J. Krecek and M. Haigh (eds) *Environmental Role of Wetlands in Headwaters*, Dordrecht, Springer Verlag, pp143–154

Castelle, A. J., Johnson, A. W. and Conolly, C. (1994) 'Wetland and stream buffer size requirements: A review', *Journal of Environmental Quality*, vol 22, pp878–882

CBT (2005a) *Colombia Basin Trust*, www.cbt.org

CBT (2005b) *Colombia Basin Trust Water Initiatives*, www.cbt.org/water

Gleick, P. H., Haasz, D. and Wolff, G. (2004a) 'Urban water conservation: A case study of residential water use in California', in Gleick, P. H. (ed.) *The World's Water 2004–2005: The Biennial Report on Freshwater Resources*, Washington DC, Island Press

Gleick, P. H., Srinivasan, V., Henges-Jeck, V. and Wolff, G. (2004b) 'Urban water conservation: A case study of commercial and industrial water use in California', in Gleick, P. H. (ed.) *The World's Water 2004–2005: The Biennial Report on Freshwater Resources*, Washington DC, Island Press

GVRD (2001) *Brunette Basin Watershed Plan*, Greater Vancouver Regional District, Vancouver, British Columbia, www.gvrd.bc.ca/sewerage/plan2001/plan.pdf

Horner, R. R., Booth, D. B., Azous, A. and May, C. W. (1997) 'Watershed dominance of ecosystem functioning', in Roesner, L. A. (ed.) *Effects of Watershed Development and Management on Aquatic Ecosystems*, New York, American Society of Civil Engineers

Khagram, S. (2004) *Dams and development: Transnational Struggles for Water and Power*, Ithaca, Cornell University Press

Labat, D., Godderis, Y., Probst, J. L. and Guyot, J. L. (2004) 'Evidence for global runoff increase related to climate warming', *Advances in Water Resources*, vol 27, pp631–642

Landcare (2005) *Integrated Catchment Management Program. Manaaki Whenua Landcare Research*, www.landcareresearch.co.nz/research/programme.asp?Proj_Collab_ID=59

Leaf, S. S., and Chatterjee, R. (1999) 'Developing a strategy on eutrophication', *Water Science and Technology*, vol 39, pp307–314

Lee, J. H., Bang, K. W., Ketchum, L. H., Choe, J. S. and Yu, M. J. (2002) 'First flush analysis of urban stormwater runoff', *The Science of the Total Environment*, vol 293, pp163–175

Mallin, M. A., Ensign, S., Wheeler, T. and Mayes, D. B. (2002) 'Pollutant removal efficacy of three wet detention ponds', *Journal of Environment Quality*, vol 31 pp654–660

Oweis, A., Hatchum, J. and Kijne, K. (1999) *Water Harvesting and Supplemental Irrigation for Improved Water Use Efficiency*, SWIM Paper No 7, Colombo, International Water Management Institute

Postel, S. and Richter, B. (2003) *Rivers for Life*, Washington DC, Island Press

República de Colombia (2004) *Proyecto de Ley del Agua*, draft version, 21 September, Bogotá, República de Colombia, Ministério de Ambiente, Vivienda y Desarrollo Territorial,

Ritter, W. F. and Shirmohammadi, A. (2001) *Agricultural Nonpoint Source Pollution*, New York, Lewis Publishers

RNRRC-Bajo (2002) *Lingmutey Chhu: Bhutan Watershed Project*, Renewable Natural Resources Research Center, Ministry of Agriculture, Royal Government of Bhutan, www.ires.ubc.ca/projects/lingmutey/html/main.htm

Rockstrom, J. (2003) 'Managing rain for the future', in Figueres, C. M., Tortajada, C. and Rockstrom, J. (eds) *Rethinking Water Management: Innovative Approaches to Contemporary Issues*, London, Earthscan

Rockstrom, J. and Gordon, L. (2001) 'Assessment of green water flow to sustain major biomes of the world: Implications for future ecohydrological landscape management', *Physics and Chemistry of the Earth (B)*, vol 26, pp843–851

Rosenberg, D. M., McCully, P. and Pringle, C. M. (2000) 'Global scale environmental effects of hydrological alterations', *BioScience*, vol 50, pp746–751

Saciwaters (2005) *Bhutan Water Policy*, www.saciwaters.org/db_bhutan_water _policy.htm#principles

Sakadevan, K. and Bavor, H. J. (1999) 'Nutrient removal mechanisms in constructed wetlands and sustainable management', *Water Science Technology*, vol 40, pp121–128

Schuler, T. (1994) 'The importance of imperviousness', *Watershed Protection Techniques*, vol 1, pp100–111

Tennant, T. L. (1975) 'Instream flow regimes for fish, wildlife recreation and related environmental resources', *Fisheries*, vol 1, pp6–10

TOL (2005a) *Water Resources Management Strategy*, Township of Langley, BC, Canada, www.tol.bc.ca/index.php?option=com_content&task=view&id=1079 &Itemid=917

TOL (2005b) *Water Wise Public Outreach Program for the Hopington Groundwater Management Area*, Township of Langley, BC, Canada, www.tol.bc.ca/index.php? option=com_content&task=view&id=1061&Itemid=897

Viviroli, D., Weingartner, R. and Messerli, B. (2003) 'Assessing the hydrological significance on the world's mountains', *Mountain Research and Development*, vol 23,
   pp32–40
von Westarp, S., Chieng, S. and Schreier, H. (2004) 'A comparison of low-cost drip irri-
   gation, conventional drip irrigation and hand watering in Nepal', *Agricultural
   Water Management*, vol 64, pp143–160
WLAP (2002) *Developing and Implementing an Integrated Stormwater Management
   Plan*, Chapter 9, Stormwater Planning Guidebook, Ministry of Land, Water and
   Air Protection, Vancouver, BC, www.env.gov.bc.ca/epd/epdpa/mpp/stormwater/
   stormwater.html
Wolff, G. and Gleick P. H. (2002) 'The soft path for water', in P. Gleick (ed.) *The
   World's Water 2002–2003: The Biennial Report on Freshwater Resources*,
   Washington DC, Island Press

6

Interdisciplinary Research and Management in the Highlands of Eastern Africa: AHI Experiences in the Usambara Mountains, Tanzania

Jeremias G. Mowo, Riziki S. Shemdoe and Ann Stroud

Brief description of project area

Location

The African Highlands Initiative (AHI) started in 1995 as a collaborative eco-regional programme focusing on natural resource management in the highlands of East and Central Africa (ECA). The programme operates under the umbrella of the Association for Strengthening Agricultural Research in East and Central Africa (ASARECA) and forms the East African component of the Global Mountain Programme (GMP) of the Consultative Group on International Agricultural Research (CGIAR). It operates as a research-for-development consortium to improve the livelihoods of poor people living in the ECA highlands. The AHI started with four countries (Madagascar, Ethiopia, Kenya and Uganda); Tanzania joined in 1997, so that five of the ten East and Central African countries are now active in the programme. The programme works in district-sized benchmark sites with a small catchment area – often a village – as the pilot site. By 1998, there were eight benchmark sites: two in Madagascar (Antsirabe and Fianarantsoa), two in Kenya (Embu and Kakamega), two in Ethiopia (Ginchi and Areka), and one each in Uganda (Kabale) and Tanzania (Lushoto). Currently, the Embu site is no longer participating in AHI activities. Figure 6.1 shows the member countries of ASARECA and the countries currently active in AHI.

The selection of benchmark sites followed these criteria: high population density (>100 people/km²), adequate rainfall (>1000mm/year) and high altitude (1400–3200m above sea level). In addition to these, there should be signs of stress, such as decreasing crops and livestock production, fragmentation of

Figure 6.1 Map of Africa showing ASARECA member countries (shaded) and countries currently participating in AHI activities (horizontal lines)

Source: chapter author

land holdings to small uneconomic sizes, increasing numbers of rural poor, limited cash opportunities, poor access to markets and limited opportunity to practise traditional ways of maintaining land productivity measures. Within the participating countries, the benchmark sites are under a host national agricultural research institute (NARI) and managed by a site coordinator who reports to the director of the host NARI, as well as to the regional AHI coordinator based in Kampala, Uganda.

Natural environment

The highlands of ECA cover about 23 per cent of the region's landmass and are characterized by steep slopes, valleys and generally undulating topography. They are fragile ecosystems liable to rapid deterioration when subjected to destructive forces such as deforestation, poor agricultural practices and overgrazing. The ECA highlands are very diverse, given a number of ecologies, histories, levels of development and socio-economic factors. Based on area, rainfall and population, Notenbaert (2004) divides the highland (altitude ≥1200m) eco-regions into four zones (see Table 6.1). Zone 1 is characterized by adequate rainfall and a high population density. In Zone 2, there are slightly fewer people but rainfall is still adequate. Zone 3 has fewer people again and less rainfall. Zone 4 receives less than 600mm per annum but has a relatively higher population density. Generally, the ECA highlands experience relatively high annual rainfall and have productive soils relative to the adjacent marginal lowlands. For these reasons, the highlands have been a major source of food and nutritional security, producing more than 50 per cent of staple food requirements in the region and several important export crops. They are sources of forest products and are important watersheds, providing water to large populations in the highlands and in lowland areas.

Table 6.1 *Zonation of highland areas in East and Central Africa based on population levels and annual rainfall*

Zone	Description	Area (km²)	Population	Density (people/km²)	Annual rainfall (mm/year)
1	Intensive highlands	313,503	73,337,815	234	≥1000
2	Drier, less intensive highlands	408,683	14,228,140	35	≥1000
3	Intensive highlands	428,030	15,891,005	37	600–1000
4	Other highlands	242,922	26,430,742	109	<600

Source: Stroud and Peden (2005)

Human settlements and activities

The ECA highlands house over 50 per cent of the regional population, most of whom (70–90 per cent) are involved in farming. Population densities are high (see Tables 6.1 and 6.2) and increasing fast. The inheritance system has led to the fragmentation of land into small parcels. Average farm size is small, ranging from 0.5 to 1.4ha per household. Given household sizes of six to eight persons, the amount of land available per person is very small (0.07–0.23ha) (Stroud and Peden, 2005). Table 6.3 shows average farm sizes in the AHI benchmark sites. As noted above, the major preoccupation of the ECA highland population is farming, most farmers being agropastoralists. Major food crops include maize, beans, bananas, wheat, barley, teff and rice; tea, coffee and horticultural products are the major export crops. Various dairy products are also produced. The highest cattle densities (103–113 head of cattle/km²) tend to occur where there are high densities of people who have a tradition of keeping livestock and where there are less intensive systems, for example, in Ginchi and Areka (Ethiopia) and Vihiga (Kenya) (Stroud and Peden, 2005). Apart from agriculture and livestock production, farmers in the ECA highlands also engage in non-farm activities including carpentry, blacksmithing, construction, stone crushing, sand harvesting, charcoal making, running small businesses like rural shops and, in a few cases, as in Lushoto (Tanzania), as tour guides for ecotourists.

Table 6.2 *National populations settled in the highlands in East and Central Africa*

Country	Total area in highlands (km²)	Total population in highlands	Density (pop/km²)
Burundi	22,629	5,467,000	242
Kenya	132,766	21,799,000	164
Madagascar	59,071	5,117,000	87
Tanzania	331,211	14,038,000	42
Eritrea	25,158	2,110,000	84
Rwanda	24,253	7,644,000	315
Ethiopia	533,825	54,409,000	102
Uganda	68,572	9,754,000	142
Sudan	15,425	185,000	12
DR Congo	180,228	9,364,000	52

Note: 'highlands' denotes >1200m above sea level.

Source: Notenbaert (2004)

Table 6.3 *Average farm sizes in the AHI benchmark sites*

Benchmark site location	Country	Average farm size (ha)	Farm land per person (ha)	Number of growing seasons per year
Emuhaya (Vihiga)	Kenya	1.0	0.125	2
Galessa PA (Ginchi)	Ethiopia	2.9	0.58	1
Gunumo PA (Areka)	Ethiopia	0.5	0.07	1–2
Kashambya (Kabale)	Uganda	1.1	0.16	2
Kianjuki (Embu)	Kenya	1.2	0.17	2
Kwalei (Lushoto)	Tanzania	1.3	0.16	2
Sahasoa (Fianarantsoa)	Madagascar	1.4	0.23	1

Source: Stroud and Peden (2005)

The African Highlands Initiative

Goals

The overall goal of the AHI is to improve the ability of grass-roots, service and support organizations and policies to achieve better food and nutritional security, income derived from agricultural activities and natural resource management (NRM) in the intensively cultivated and densely populated highlands of ECA. This goal will be achieved through inter-institutional and multidisciplinary research and development (R&D) efforts with strong community participation. The main purpose of AHI is to develop methods and approaches for integrated natural resource management (INRM) and to institutionalize these for improved effectiveness and efficiency of R&D efforts.

Specific objectives

The specific objectives of AHI are:

- to develop and test approaches and methodologies to improve the effectiveness of R&D;
- to develop and implement district-based strategies to improve the orientation of development, policy interventions and practices;
- to induce change in research institutional policy and practice for enhanced use of INRM approaches;
- to collect, organize and disseminate information, methodologies and technologies to R&D agencies and farmers;
- to mobilize and strengthen partnerships and links between institutions to address agroecosystems and NRM issues more efficiently.

Definition of objectives

In 1992, representatives of NARIs in the ECA region and the CGIAR centres (referred to below as the founder group) met in Nairobi, Kenya, and proposed the formation of an eco-regional programme catering for the highlands of ECA. The major task of the programme would be to integrate expertise and perspectives from different organizations to address complex socio-economic and technical problems that had not been solved through conventional R&D approaches. The formation of AHI was prompted by the fact that agricultural research conducted for many years in the high-potential and densely populated highlands of the region had not achieved adequate results in terms of improved and sustainable productivity. It was therefore felt that there should be a change in the way research was conducted and the way that R&D organizations related to each other.

A two-person consultant team was commissioned to conduct a thorough study, which was presented to the founder group. The recommendations and focus suggested were accepted. A regional stakeholders' workshop was held to provide further direction, and in September 1995 AHI was established as both a CGIAR and an ASARECA programme. Subsequently, a task force and, later, a regional steering committee were formed to provide guidance on the technical direction of AHI and governance. Members were institutional representatives from NARIs, international agricultural research centres (IARCs) and donors. Several regional technical committees that took various shapes over time and worked on the technical focus and operational details augmented this group. Various regional, well-facilitated stakeholder meetings that formulated and updated ideas for direction were held. A consultative process was followed throughout.

Groups involved

The AHI was developed with the full participation of the NARIs in ECA and the IARCs active in the region, with the full endorsement of ASARECA. NARIs present at the inception of the programme in 1995 included those from Kenya, Uganda, Madagascar and Ethiopia. In 1997, Tanzania joined the programme. Recently, NARIs from Rwanda, Burundi, Eritrea and the Democratic Republic of Congo have expressed interest in joining. The IARCs active in the region include the Centro Internacional de Agricultura Tropical (CIAT), Centro Internacional de la Papa (CIP), Centro Internacional de Mejaramiento de Maiz y Trigo (CIMMYT), International Institute for Tropical Agriculture (IITA) and International Livestock Research Institute (ILRI). Other active organizations are the International Food Policy Research Institute (IFPRI), International Centre for Research on Agroforestry (ICRAF), International Plant Genetic Resources Institute (IPGRI), International Centre for Insect Physiology and Ecology (ICIPE) and the International Crop Research Institute for the Semi-Arid Tropics (ICRISAT).

Time-frame

The programme runs in phases of three years. Phase 1 started in January 1995 and ended in December 1997. In this phase, there was a disciplinary, researcher-oriented agenda, with individual scientists working in geographically scattered locations, largely supported by small grants. The second phase started in 1998 and ended in 2000. As it was felt that the approach used in Phase 1 was not effective in fostering an INRM agenda, in Phase 2 it was proposed that the AHI move towards a geographically concentrated (pilot sites), team-based, farmer-driven agenda focusing on NRM and productivity issues (AHI, 2001). One of these pilot sites was the Lushoto benchmark site and work at this site started in Phase 2. The year 2001 was a transition one, when time and resources were devoted to critically examining whether AHI goals were being achieved. During Phase 3 (2002–2004), the major thrust was integrated watershed management, capitalizing on the results obtained from the pilot sites to reach more farmers in the benchmark sites and beyond.

Funding sources

Phase 1 (1995–1997) was supported as a start-up phase by The Netherlands government, the Swiss Agency for Development and Cooperation (SDC) and the International Development Research Centre (IDRC). The US Agency for International Development (USAID) and the Rockefeller Foundation also provided funding. Many organizations and institutions, participating NARIs, IARCs, government organizations and non-governmental organizations (NGOs) contributed staff time and material resources for the implementation of activities. During Phases 2 (1998–2000) and 3 (2002–2004), in addition to the above sources, AHI received support from the Consortium for the Sustainable Development of the Andean Ecoregion (CONDESAN), the UK Department for International Development (DfID) and the Ford Foundation.

AHI experiences in the Usambara Mountains, Tanzania

Location characteristics

The AHI benchmark site in Tanzania is Lushoto District in the northeastern part of the country. About 80 per cent of the district is occupied by the West Usambara Mountains (see Figure 6.2). These form part of the Eastern Arc Mountains and rise to 900–2250m. With a population of 418,652 (URT, 2002) in an area of 3500 km², Lushoto is one of the most densely populated districts in the country. The importance of the West Usambara Mountains, and efforts to conserve them, has received much coverage (Iversen, 1991; Johansson 2001; Pfeiffer, 1990; Stroud, 2000). Their importance is not only due to the high agricultural potential but also because they are a source of various products to populations and industries within and outside the mountains, including water for irrigation and hydropower generation (Iddi, 2000). The Department of Research and Development (DRD) of the Ministry of Agriculture and Food Security chose Lushoto as the AHI benchmark site because the area fits the general criteria and has several active organizations that are key in forging partnerships to accomplish the AHI goals. Researchers of different backgrounds and experience come from four institutes and two NGOs to form the AHI Lushoto site team.

Major NRM constraints in the West Usambara Mountains

The major problems hindering sustainable NRM in the West Usambara Mountains include overpopulation, land degradation, excessive deforestation, poor market infrastructure and poorly formulated policies and by-laws (Mowo et al, 2002a). Land scarcity is a major problem, leading to widespread cultivation of marginal and restricted lands. Excessive deforestation has led to declining amounts and reliability of rainfall (Huwe, 1988), declining water levels, increased incidences of heavy runoff and floods (Mowo et al, 2002b), and loss of biodiversity (Newmark, 1998). Poor land and livestock husbandry has led to declining yields, increased food insecurity, poor nutrition and increased dependence on forest resources for livelihoods.

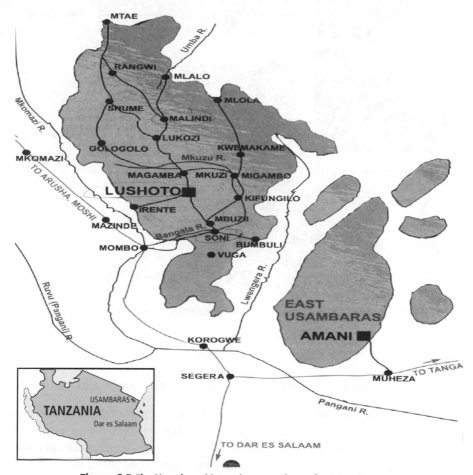

Figure 6.2 The Usambara Mountain ranges in northeastern Tanzania

Poor access to technological and market information has led to non-use of available technologies for, and limited investment in, NRM by farmers. Meanwhile, poorly formulated policies have hindered smooth implementation of sustainable practices of NRM. Development of such policies did not always involve all stakeholders – notably the farmers – and they are not clear or sufficient with regard to aspects of NRM.

Interdisciplinarity in the project team

Composition of the research team

Figure 6.3 shows the composition of the research team in Lushoto. There is a strong predominance of biophysical scientists (86 per cent), compared to the other disciplines. Social scientists are the least represented. This scenario is more or less the same in the other AHI sites.

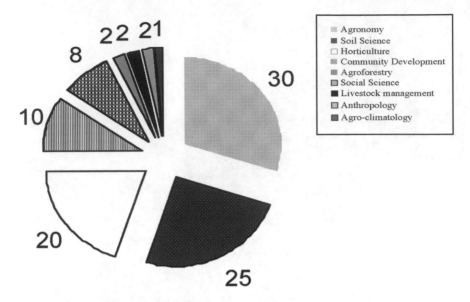

Figure 6.3 Distribution of the different professionals in the Lushoto AHI site team

Mechanisms used to foster interdisciplinary concepts and working

To foster interdisciplinary concepts and working, we held meetings involving professionals from the relevant disciplines, conducted joint identification of important entry points to the community, facilitated the development of joint project proposals in reaction to constraints identified by farmers, and conducted interdisciplinary visits to the community. The AHI had a regional research fellow specifically working on setting up performance self-monitoring to evaluate progress in interdisciplinarity, teamwork, participatory research and use of partnerships (linkages). Individual and group reflection on successes and barriers helped to make researchers more aware of the benefits of such an approach. The work centres on problem orientation, described in the form of research questions or hypotheses. A single discipline or a commodity approach cannot solve most of these questions. We have attempted to bring in farming systems or landscape perspectives to the research, needing an integrated approach. This has been a struggle in terms of expertise because most researchers are oriented towards working on single components rather than systems or landscapes. We have also used two 'protocol' guides, developed by AHI headquarter staff with inputs from the sites, to help scientists define their research in an interdisciplinary way. The guides consider empirical or formal research, and process or action research. To date, they are unpublished.

Under *empirical* or *formal research*, scientists are able to delineate a research design from the outset, and define a relatively fixed research framework that is designed to match the specific question at hand. Often these are the 'what' questions normally used by biophysical or social scientists, for example, what are the major causes of 'x'? What are the major types of 'y' existing in the community? Empirical research questions are used not only for biophysical but also for social science research. In the latter, comparisons (and corresponding 'replicates') encompass people or groups of people representing distinctive social, cultural or political circumstances (for example, different social categories, ethnic groups or livelihood systems).

Process or action research is a more flexible research framework and often it is the 'how' questions guiding the action research, for example, how do we best do 'x'? The research findings only emerge as approaches are tested and improved upon. Action research diverges a great deal from that usually used by biophysical scientists. Due to its iterative and experiential nature (an ongoing process of planning → implementation → monitoring → modification → implementation), process research is best combined with process documentation and/or formal monitoring and evaluation systems. This is because, unless learning and reflection occur, it will be difficult to learn from mistakes and to use this learning to streamline the approach, and to know whether the approach is effective from the standpoint of diverse social actors. While it is possible to follow certain principles of good facilitation or process reflection, these skills are best developed through an iterative process of practice and analysis.

Challenges, successes and failures – and why

Challenges
The Lushoto team has encountered several challenges to interdisciplinary work including:

- lack of mutual understanding, respect and appreciation due to varying interests, ambitions and expectations among different professionals;
- limited team spirit, evidenced by the reluctance of some professionals to work together, due probably to lack of confidence, difficulty for some professionals to change from previous ways of doing things, and a reward system that does not recognize individual efforts in a team;
- logistical constraints, mainly in coordinating professionals from different institutes who have other research duties. Some AHI sites are far from the institutes where participants are based, making travel and subsistence costs high.

Successes
There is a growing appreciation by biophysical scientists of the need for other disciplines to contribute to solving the complex issues of NRM. Most professionals are gradually accepting the approach of working in a team as they

spend more time with each other and with farmers in the field. Our partners are appreciating the importance of sharing knowledge and experience and are, in turn, being enriched by the farmers' vast indigenous knowledge. Comments by a senior plant breeder in the AHI team in Lushoto tell it all:

> *I never thought farmers have anything to do with conventional plant breeding. Neither did I consider other disciplines as important as mine. After working with farmers and other professionals, I now know why some of our new releases were not popular or received by our farmers. I am also convinced that no one profession can address farmers' constraints adequately. Each one of us including the farmers has something important to contribute to the R&D process. In other words we need each other and we need therefore to move towards and interact with each other.*

These comments are in line with the observation of Chambers (1980) that 'it is fascinating and illuminating to work in field situations with people from other disciplines'. We have learned that interdisciplinary research requires patience among the participating professionals. At the beginning, trying to build an understanding among ourselves was time-consuming but eventually rewarding.

Failures

It was the intention of the project to invest in capacity building of the different professionals working at the site, to enable them to acquire the skills required for effective interdisciplinary work. However, financial constraints have limited such investments. Furthermore, we have not invested enough in scaling up the approach so that our research system can be implemented at the national level. Recommendations to make interdisciplinary research more effective include the need to build capacity in the less-represented disciplines, to expose teams to successful cases elsewhere, and for NARIs in the region to develop a reward system that recognizes individual efforts in team work.

Stakeholder involvement

Definition of issues, priorities and activities

Issues, priorities and activities are defined in different ways at the four levels with which the project is concerned: community, district, national and regional levels. At the community and district levels, we use participatory rural appraisal (PRA) techniques, focus group discussions and key informant interviews (with researchers, extension staff, NGOs, farmers and local government). At the national level, we hold consultations with major implementing organizations to ensure that regional and community agendas fit into their priorities and plans, and to get their commitment (from researchers and NGOs and their managers).

At the regional level (ECA), we use stakeholder meetings and a technical committee to make decisions that are more specific. We also have a steering committee involving representatives of national and international research organizations with the main task of overseeing and offering advice on the implementation of the AHI mandate.

Implementation of research

At the community or pilot site level, we work directly with farmers, community groups, community facilitators, NGOs and extension staff. At the regional (ECA) level, we have a full-time research team with skills that are missing in the site research teams (anthropology, sociology and development economics) who mentor and conduct some additional research in collaboration with the biophysical scientists.

Results and outcomes of the project

Technologies

Several technologies that suit various ecologies in highland areas have been developed and tested using a participatory technology development (PTD) approach. These include technologies on integrated soil fertility management (Gachene et al, 1999; Wickama and Mowo, 2001), integrated pest and disease management (Minja et al, 2003), soil and water conservation, and multipurpose trees and legume varieties and their management (AHI, 2001).

Methodologies, tools and approaches

Different methodologies, tools and approaches have been developed including those for improving systems and landscapes, improving R&D strategies, and improving institutional capacity to achieve sustainable development (Amede, 2003; German, 2003a, 2003b; German et al, 2004; Stroud, 2002, 2003a, 2003b, 2003c, 2003d; Stroud and Khandelwal, 2003).

The methodologies, tools and approaches for improving systems and landscapes aim at:

- improving the ability of farmers and natural resources managers to innovate and integrate technology and management options at farm and landscape levels;
- supporting combined science and farmer-led research on technology options to develop useful, adapted technologies that fit into local ecologies and social settings (PTD approach);
- facilitating social dynamics at community level to solve technical issues, improve collective action, conflict resolution and landscape management;

- facilitating institutional arrangements and platforms at local, district and national levels for improved joint analysis and implementation of development plans and to influence policy (community lobbying and local policy formulation for improving NRM collective action); and
- influencing policy-making and policy-makers using a grass-roots approach.

The methodologies, tools and approaches for improving R&D strategies focused on the development of communication strategies to improve information provision, expansion of strategies that link market and policy support, and provision of incentives to community action for improved investment in natural resource regeneration and maintenance. Others include: improving methods for targeting development and NRM agendas and initiatives to be more inclusive of the resource-poor, HIV/Aids-affected and other marginalized groups; exploring and communicating local and broader impacts of technologies, methods and processes enabling development, and policies on environment, equity and livelihoods; and influencing the agendas and methods used by support and service organizations aimed at development.

The methodologies, tools and approaches for improving institutional capacity to achieve sustainable development aim at facilitating research organization managers to support research for development and increase impact through expanded agendas and partnerships.

Participatory monitoring and evaluation

Participatory monitoring and evaluation (PM&E) during Phase 2 of the AHI at the benchmark site in Lushoto assessed the extent to which site teams were incorporating interdisciplinarity, partnership and participatory research in their work. Results showed that scientists were strong in conducting interdisciplinary research and were able to give many examples of how it improved the teams' effectiveness in solving farmers' problems. With regard to linkages, the team had a strong working relation with the Ministry of Agriculture and some non-governmental and community-based organizations. Links with policy-makers were weak because most scientists did not see them as important to their work. However, this has changed in subsequent phases because we are operating at the larger scale of the watershed level. Because links with community-based organizations, community leaders and policy-makers are important for successful watershed intervention, deliberate efforts are made to have them play key roles, such as mobilization of communities and taking a leading role in NRM.

Scientists noted that they were strong in some areas of participatory research, such as working with farmers' research groups (FRGs), implementation of trials and in sharing results with farmers, but they were weak in designing and evaluating trials with farmers. It was felt that there was a need to monitor more systematically the effects and outcomes of the use of these approaches by researchers. For example, as researchers adopt interdisciplinary research, what are the specific changes that take place in their behaviour, relationships, activities and actions? Using outcome mapping to monitor these effects will provide important lessons on the performance of researchers, including an assessment of the progress they are making in moving towards desired changes.

Dissemination of results and outcomes

Results and outcomes are disseminated at various levels. At the local level, the AHI R&D teams make an effort to disseminate technologies through the extension system and encourage farmer-to-farmer transfer. Methodologies are disseminated to NGOs, extension and researcher workers through working together in the field, publications (for example, Barrios et al, 2002), field days and demonstrations. Various actors participate in the dissemination of results and outcomes. They include farmers, community leaders, extension workers, researchers and NGOs. Through technology spillover, farmers have participated fully in dissemination of the results. This happens when farmers pass through a field and see the performance of the various technologies that have been adopted by their colleagues directly involved in the project. They stop and ask questions that may finally lead to a request for seed, or for an explanation on how to conserve and fertilize soils. Through this pathway, technologies introduced at the pilot site have spread to distant places in Lushoto district. There is also spontaneous dissemination taking place in the benchmark sites. For example, in Lushoto we found that the most preferred technologies, such as improved tomato, cabbage and banana germplasm, are spreading with no intervention from the extension department. This kind of technology spillover is mainly achieved through friends, old schoolmates, colleagues and family and kin relations (German et al, 2004). A high concentration of spillover is found within the pilot site, followed by adjacent villages.

Community leaders, including leaders of the village governments, councillors and religious leaders, participate in disseminating results during sensitization sessions and indirectly by acting as a model in their communities through having better-managed fields. Extension workers, NGOs and researchers working with the farmers also promote technology uptake through sensitization sessions, preparation of leaflets and manuals, and conducting tours for farmers from other villages to the project site.

For the scientific communities and other stakeholders at national, regional and global levels, we conduct and attend workshops and we have AHI Briefs that summarize major findings from the sites (for example, Amede, 2003; German, 2003a, 2003b; German et al, 2003; Stroud, 2003a, 2003b). In October 2004, a regional AHI workshop was held in Nairobi, entitled 'Integrated Natural Resource Management in Practice: Enabling Communities to improve Mountain Livelihoods'. The workshop attracted participants from different countries and organizations interested in the management of natural resources in mountain ecosystems. Researchers from the Lushoto benchmark site presented several papers (for example, German et al, 2004; Lyamchai et al, 2004a, 2004b; Mansoor et al, 2004; Meliyo et al, 2004; Mowo et al, 2004a, 2004b, 2004c, 2004d; Ngatunga et al, 2004; Owenya et al, 2004). In another workshop held in Dar es Salaam, Tanzania, a paper on farmers' perception on eucalyptus was presented (Mowo et al, 2005). Papers have also been given at international meetings, for example, a symposium in Wageningen (The Netherlands) on 'Nutrient Management in Tropical Agro-ecosystems' in February 2004 (Mowo et al, 2004d). Global dissemination is through the AHI website (www.africanhighlands.org) and AHI annual reports (for example, AHI, 2001).

Implications of the project

The project has influenced decisions and policies at a number of levels. At the local level in pilot villages, AHI committees have been established and accepted as part of the village administrative structure. The AHI committee is represented in the village government meetings and plays an active role in pushing for the adoption of technological options identified and verified in the area. At a wider regional scale, AHI has reviewed policies related to NRM and offered suggestions to relevant local authorities (for example, ward and district councils). While we have not yet been able to influence national policies on NRM, one of our current and future thrusts is to convince relevant authorities and institutions at national level of the need to adopt the AHI approach. This will be achieved through demonstration of the effectiveness of our approach. Finally, projects working in partnership with AHI have, in principle, accepted the need for a collective approach in addressing farmers' constraints. Consequently, they have asked for AHI participation in their activities. One example is the collaborative Beyond Agricultural Productivity to Poverty Alleviation/Traditional Irrigation Project (BAPPA/TIP) in Lushoto.

Lessons for integrated research and management

The major focus of AHI is INRM for sustainable system productivity in the highland ecosystems of ECA. This is a complex undertaking, given that natural resource management requires not only biophysical solutions but also consideration of social, economic, policy, marketing and infrastructure aspects. Consequently, integrated research calls for different disciplines and stakeholders to work together to holistically address constraints in a way that yields broad results/outcomes to solve community problems. This also implies that communities living in the mountains should be key actors in INRM research. Consequently, the major challenges to successful integrated research in NRM will be the identification of strategic partners, building interdisciplinary team spirit, understanding farmers' needs, and making project goals meet these needs. Other challenges include building the culture of interdisciplinarity, whereby team members are patient with each other as each one tries to understand the other's profession and perspectives. Building commitment, trust and a learning culture among team members to acquire farmers' indigenous knowledge and to understand and respect their point of view is another challenge to effective integrated research in NRM. Finally, the current reward system in most countries does not promote integrated research because recognition of effort is based on individual rather than collective achievement.

To overcome these challenges requires time to change our ways. A reward system based on collective achievement should be put in place, while short courses should be conducted to empower professionals in integrated research. More funds should be allocated for research, with emphasis on long-term projects, since it will take time before the fruits of integrated research are evident.

References

AHI (African Highland Initiative) (2001) *1998 Annual Report*, Kampala, African Highland Initiative

Amede, T. (2003) 'Differential entry points to address complex natural resource constraints in the highlands of Eastern Africa', in *Managing Farms and Landscapes*, AHI Brief C1, November, Kampala, African Highland Initiative

Barrios, E., Bekunda, M., Delve, R., Esilaba, A. and Mowo, J. (2002) *Identifying Local Soil Quality Indicators: Methodologies for Decision Making in Natural Resource Management, Eastern African Version*, Cali, International Center for Tropical Agriculture

Chambers, R. (1980) *Understanding Professionals: Small Farmers and Scientists*, New York, International Agricultural Development Service

Gachene, C. K. K., Palm, C. A. and Mureithi, J. G. (1999) *Legume Cover Crops for Soil Fertility Improvement in the Eastern Africa Region*, Report of an AHI workshop 18–19 February, Nairobi, TSBF

German, L. (2003a) 'Beyond the farm: A new look at the livelihood constraints in the highlands of Eastern Africa', in *Managing Farms and Landscapes*, AHI Brief A2, November 2003, Kampala, African Highland Initiative

German, L. (2003b) 'Watershed entry: A socially-optimal approach', in *Managing Farms and Landscapes*, AHI Brief C2, November, Kampala, African Highland Initiative

German, L., Stroud, A. and Obin, E. (2003) 'A coalition for enabling demand driven development in Kabale District, Uganda', in *Managing Farms and Landscapes*, AHI Brief B1, November, Kampala, African Highland Initiative

German, L., Kingamkono, M. and Mowo, J. G. (2004) *Scaling Up Strategically: A Methodology for Understanding Patterns and Limits to 'Spillover' of Farm-Level Innovations*, presented at the Regional AHI Conference on 'Integrated Natural Resource Management in Practice: Enabling Communities to Improve Mountain Livelihoods', 12–15 October, Nairobi, African Highlands Initiative

Huwe, C. (1988) *Half-Yearly Report of the SECAP Research Department*, West Usambara Mountains, Lushoto, SECAP

Iddi, S. (2000) 'Eastern Arc Mountains and their national and global importance', *Journal of East African Natural History*, vol 87, pp19–26

Iversen, S. T. (1991) *The Usambara Mountains, NE Tanzania: History, Vegetation and Conservation*, Uppsala, Uppsala University

Johansson, L. (2001) *Ten Million Trees Later: Land Use Change in the West Usambara Mountains*, The Soil Erosion and Agroforestry Project in Lushoto District 1981–2000, GTZ, Eschborn, Sweden

Lyamchai, C. J., Mowo, J. G. and Wickama, J. M. (2004a) *Managing New Working Relationships: Partnership and Metworking*, Paper presented at the Regional AHI Conference on 'Integrated Natural Resource Management in Practice: Enabling Communities to Improve Mountain Livelihoods, 12–15 October, Nairobi, African Highlands Initiative

Lyamchai, C. J., Kingamkono, M., German, L., Shemdoe, R. and Mowo, J. G. (2004b) *Farmer Innovations in Natural Resource Management: Lessons and Challenges*, Paper presented at the Regional AHI Conference on 'Integrated Natural Resource Management in Practice: Enabling Communities to Improve Mountain Livelihoods', 12–15 October, Nairobi, African Highlands Initiative

Mansoor, H. A., Wickama, J. M. and Mowo, J. G. (2004) *Water Resources Management in the Baga Watershed: Past, Present and Future*, Paper presented at the Regional AHI Conference on 'Integrated Natural Resource Management in Practice: Enabling Communities to Improve Mountain Livelihoods', 12–15 October, Nairobi, African Highlands Initiative

Meliyo, J. L., Masuki, K. F. G. and Mowo, J. G. (2004) *Integrated Natural Resource Management as a Way Forward to Food Security and Poverty Eradication: The Case of Kwlaei Village, Lushoto District Tanzania*, Paper presented at the Regional AHI Conference on 'Integrated Natural Resource Management in Practice: Enabling Communities to Improve Mountain Livelihoods', 12–15 October, Nairobi, African Highlands Initiative

Minja, E. M., Ampofo, J. K. O., Mowo, J. G. and Mziray, H. A. (2003) *Training of Bean Farmers and Village Extension Officers, Lushoto District Tanga Region, Northeastern Tanzania*, Integrated Pest Management Training Workshop Report, February

Mowo, J. G., Mwihomeke, S. T., Mzoo, J. B. and Msangi, T. H. (2002a) *Managing Natural Resources in the West Usambara Mountains: A Glimmer of Hope in the Horizon*, Paper presented at the Mountains High Summit Conference for Africa, 6–10 May, Nairobi

Mowo, J. G., Tenge, A. M., Wickama, J. M., Malilo, E. R. and Shebughe, R. H. (2002b) *Alternatives for Improving Livelihoods in the Usambara Mountains*, Paper presented at the 6th International Conference on Sustainable Mountain Development in Africa: Agenda for Action, 19–24 August, Moshi

Mowo, J. G. and Shemdoe, R. S. (2004a) *Bridges of Convenience: Linking Partners for Effective Natural Resource Management*, Paper presented to the Symposium on Nutrient Management in Tropical Agro-ecosystems, 18–19 February, Wageningen

Mowo, J. G., German, L., Getachew, A. F., Mbakaya, D. S., Endrias, G. and Tolera, T. (2004b) *Local Institutions and their Role in Natural Resource Management in the East African Highlands*, Paper presented at the Regional AHI Conference on 'Integrated Natural Resource Management in Practice: Enabling Communities to Improve Mountain Livelihoods', 12–15 October, Nairobi, African Highlands Initiative

Mowo, J. G., Shemdoe, R. S., Lyamchai, C. J. and Sellungato, M. (2004c) *Attitude Change for Effective Natural Resources Management and Development: Who Should Change?*, Paper presented at the Regional AHI Conference on 'Integrated Natural Resource Management in Practice: Enabling Communities to Improve Mountain Livelihoods', 12–15 October, Nairobi, African Highlands Initiative

Mowo, J. G., Lyamchai, C. J., Stroud, A. and Opondo, C. (2004d) *The Challenges and Opportunities in Leading a Multi-Disciplinary Team of Professionals from Multiple Institutions: Lessons from AHI in East Africa*, Paper presented at the Regional AHI Conference on 'Integrated Natural Resource Management in Practice: Enabling

Communities to Improve Mountain Livelihoods, 12–15 October, Nairobi, African Highlands Initiative

Mowo, J. G., Wickama, J. W. and Mathias, S. (2005) *Eucalyptus and Water Sources in the Usambara Mountains: Farmers' Perception*, Paper presented at the Sensitization Workshop on Eucalyptus, 17–18 February Dar es Salaam, Tanzania

Newmark, W. D. (1998) 'Forest area, fragmentation and loss in the Eastern Arc Mountains: Implications for conservation of biological diversity', *Journal of East African Natural History*, vol 87, pp29–36

Ngatunga, E. L., Lema, N. M. and Mowo, J. G. (2004) *Institutionalisation of the AHI's Inter-Disciplinary and Multi-Institutional Approach into the DRD System in Tanzania*, Paper presented at the Regional AHI Conference on 'Integrated Natural Resource Management in Practice: Enabling Communities to Improve Mountain Livelihoods', 12–15 October, Nairobi, African Highlands Initiative

Notenbaert, A. (2004) *AHI Zone Map*, Nairobi, International Livestock Research Institute

Owenya, M., Mowo, J. G., Shemdoe, R. and Matosho, G. (2004) *Traditional Dances for Technology Dissemination*, Paper presented at the Regional AHI Conference on 'Integrated Natural Resource Management in Practice: Enabling Communities to Improve Mountain Livelihoods', 12–15 October, Nairobi, African Highlands Initiative

Pfeiffer, R. (1990) *Sustainable Agriculture in Practice: The Production Potential and Environmental Effects on Macro-contour lines in the West Usambara Mountains of Tanzania*, PhD Dissertation, Hohenheim, University of Hohenheim

Stroud, A. (2000) *African Highlands Initiative (AHI) Progress Report*, Kampala, African Highlands Initiative

Stroud, A. (2002) *Regional Synthesis: Operationalizing AHI Phase III – Developing Approaches for Improving Integrated Watershed Management*, AHI Working Paper 14, Kampala, African Highlands Initiative

Stroud, A. (2003a) 'Linked technologies for increasing adoption and impact', in *Managing Farms and Landscapes*, AHI Brief A3, November, Kampala, African Highlands Initiative

Stroud, A. (2003b) 'Combining science with participation: Learning locally and generalizing regionally', in *Managing Farms and Landscapes*, AHI Brief C3, November, Kampala, African Highlands Initiative

Stroud, A. (2003c) *AHI Profile Report to ASARECA: Broadening Horizons – Institutional, Policy and Technical Innovations for Improving NRM and Agricultural Productivity in the East and Central African Highlands (2002–2004)*, AHI Working Paper 16, Kampala, African Highlands Initative

Stroud, A. (2003d) 'Transforming institutions to achieve innovation in research and development', in Pound, B., Snapp, S., McDougall, C. and Braun, A. (eds) *Managing Natural Resources for Sustainable Livelihoods: Uniting Science and Participation*, London, Earthscan

Stroud, A. and Khandelwal, R. (2003) *In Search of Substance: An Analytical Review of Concepts and Approaches in Natural Resource Management in Sub-Saharan Africa*, AHI Working Paper 20, Kampala, African Highlands Initiative

Stroud A. and Peden, D. (2005) *Situation Analysis for the Intensity Cultivated Highlands of East and Central Africa Part A: 2005; An Input into the AHI strategy for ASARECA 2005 – 2010*, Kampala, African Highlands Initiative

URT (United Republic of Tanzania) (2002) *Population and Housing Census Volume II. Age and Sex Distribution*, Central Census Office, Dar es Salaam, National Bureau of Statistics

Wickama, J. M. and Mowo, J. G. (2001) *Indigenous Nutrient Resources in Tanzania*, Managing African Soils, no 22, Edinburgh, Drylands Programme, IIED

7

Interdisciplinary Research and Management in Mountain Areas of Arumeru District, Northern Tanzania

Fidelis B. S. Kaihura

Introduction

For thousands of years, farmers have constantly modified their use and cultivation of biodiversity for food and livelihoods through learning, experiment and innovation. They have nurtured and managed a diversity of plants and animals – both wild and domesticated – and developed agrodiversity to harness various plants and animals for human benefit. In this process, types of agricultural land use have diverged. High-yielding varieties have often replaced the huge diversity of local varieties and genotypes. However, for several reasons, pockets of small-scale agriculture and land use have remained. Often, such land use is more appropriate to feed high densities of rural populations. One example is provided by the intensive home gardens, represented classically in northern Tanzania by the Chaga home gardens, also known as the Kibanja system, in Bukoba, northwestern Tanzania (Kaihura, 1999; Stocking et al, 2003).

The project described here was part of the global People, Land Management and Environmental Change (PLEC) project, supported by the United Nations University and the Global Environment Facility. PLEC has involved local farmers and scientists in setting up demonstration sites in critical ecosystems and areas of globally significant biodiversity, such as forest, mountain, semi-arid, freshwater and wetland, in major regions in Africa, Asia and the Pacific and tropical Americas (Brookfield et al, 2003).

In Tanzania, the PLEC project concentrated on the pockets of mainly small-scale, intensive and diverse agricultural systems on the premise that they are under threat and worth conserving. They contain value in their biodiversity and wealth of knowledge: aspects that perhaps have less resonance than food security and large-scale grain storage, but which potentially have implications for the sustainability of both large-scale commercial agriculture and small-scale diverse agriculture (Stocking et al, 2003). The general project objective was to develop sustainable resources management technologies in agricultural systems that enhance and improve biodiversity, and improve smallholder livelihoods and food security. In implementing such diverse land and plant management systems, multidisciplinary teams were needed to effectively address and achieve project objectives.

Project sites

The project area is on the slopes of Mount Meru (summit 4562m) in northern Tanzania. The mountain is volcanic; past eruptions have resulted in the formation of several cones of various sizes around the mountain. The slopes of the cones are steep and dissected, with broad U-shaped and V-shaped valleys. Olgilai and Ng'iresi villages are located on the slopes of Kivesi hill on the mid-slopes of the mountain, with some farms on steep slopes of up to 30–50 degrees but also with broad valleys. Kiserian village is on the undulating plains of the semi-arid lower slopes. The bedrock is late cretaceous to recent volcanic materials: basalt, trachytes and pyroclastics (Morss, 1980). The soils are generally rich in nutrients but need external inputs for better yields. The rainfall pattern is bimodal, with long rains from March to May, and short rains from November to December. The average annual rainfall is about 2000mm in the sub-humid zone and 500mm in the semi-arid zone. Temperatures in both zones range between 12 and 30°C. The major land-use system for the sub-humid zone is agroforestry, and in the semi-arid zone, mixed cropping. Major crops include bananas, coffee, maize, beans, irish and sweet potatoes, pigeon peas and vegetables (Kaitaba et al, 1999). The population in 2002 was about 516,814, with a growth rate of around 4 per cent. This rate is one of the highest in the country, above the national growth rate of about 2.9 per cent. The resultant population density is also high: 110 persons per km^2 (Bureau of Statistics, 1998).

Planning and implementing the project

Establishment of objectives at the local level

While PLEC was a global project, it provided for individual countries or clusters to develop local objectives. Pilot participatory rural appraisal (PRA) involving farmers, researchers, extension staff and other stakeholders in natural resources management and government institutions in the area resulted in specific objectives within the general project objective. The stakeholders also developed focused activities relevant to the Tanzanian situation. Involved groups included farmers (both men and women), extension staff, district council representatives, research and training institutions, decision- and policy-makers at community and district level, and NGOs working on biodi-versity/environmental conservation in the area.

Different groups of local stakeholders were involved in various ways in the definition of issues, priorities and activities. Farmers identified and prioritized the research issues, and expert farmers with good resource management models were identified and requested to allocate part of their land for demon-stration/experimentation. Staff involved in other projects in the area assisted in identifying areas and farmers with the resources to implement the project successfully. However, careful discussions between scientists and farmers brought in many different farmers from those identified by existing projects. Most of the researchers were from outside the project area. They facilitated farmer discussions during preliminary project meetings, identified able expert farmers to impart knowledge to other farmers, identified indigenous technolo-gies with potential for improvement and up-scaling, consulted and convinced poor farmers to join project work (as most of them were normally left out by other projects because of their poverty), and sensitized women to be proactive and join the project. The extension staff, all of whom had been working in the area for over five years, helped provide in-depth information about cultural behaviour, norms and traditions of farmers, potential production constraints and existing alternative solutions, and the interests and behaviour of current leaders at village to district level. They also introduced outside members of the project to farmers and leaders. Finally, through their active participation in project activities, village, community and district-level leaders were involved in sensitizing farmers, particularly laggards, to join project activities.

PLEC emphasized looking around in farmers' fields and identifying specific successful management practices and technologies as entry points. The willingness of farmers with successful management models to teach other farm-ers using their farms as demonstration/training sites was also looked at. This approach focused on exploiting knowledge, techniques and practices from within existing farm management technologies and on involving all categories of farmers, including the very poor. The poor are normally left out in project implementation, but are the ones who most degrade the land by exploiting nature to earn their living (Stocking, 2003).

Priority issues identified through PRA included control of land degrada-tion, enhancement and conservation of biodiversity and the environment, increased income, and capacity building of farmers in managing and promot-ing biodiversity and environmental protection. To address these issues, activities included:

- farmer-to-farmer training on expert farmers' fields;
- farmer exchange visits;
- farmer-led experiments;
- monitoring of farmer-initiated experiments;
- training of farmers to establish and manage low capital investment projects

that enhance biodiversity while improving household income;
- documentation and monitoring of changes in biodiversity over time;
- field days and workshops for feedback to stakeholders and forward planning;
- farmer-focused group activities such as rehabilitation of water sources, erosion control on steep slope farms, rehabilitation of degraded grazing lands, and management of indigenous woodlots.

Interdisciplinarity in the project team

Through the diverse activities outlined above, scientists and extension staff were also trained in the best approaches to successfully work with farmers. As both the farmers' demands and the project approach, with a focus on farming systems and landscape, necessitated interdisciplinary operations, a multidisciplinary team of scientists, extension staff and expert farmers was established to ensure successful implementation. Thus, both planning and implementation included all categories of farmers and a multidisciplinary team of professionals and actors with different backgrounds and experiences, brought together to interact in PLEC demonstration sites. Overall, the professions and proportions of scientists, extension staff and others were: soil science and agronomy (36 per cent), forestry (7 per cent), demography (4 per cent), livestock management (11 per cent), botany (11 per cent), extension (21 per cent), community development (3 per cent) and community leaders (7 per cent). While anthropologists and socio-economists were needed to complete the teams, they were difficult to find.

Staff moved as groups of combined disciplines in most cases. In addition to scientific discipline, gender balance was also addressed in the groups. The group combinations changed according to the issues being addressed. Where experts were not available, for example, for agrometeorology, external contractors were brought in. Influential farmers/leaders were included in each group.

Stakeholder involvement in implementing research

As in the definition of project objectives, priorities and activities, the different stakeholder groups played specific roles in implementing research within the project. Farmers managed research and demonstration fields, planned and conducted on-farm training of fellow farmers, and organized themselves into common interest groups for biodiversity and household income-generating activities, for example, keeping of bees for honey, protecting conserved areas against fires and tree cutting, and keeping local chickens for eggs, meat and selling. Researchers facilitated farmer-led activities of experimentation, demonstration and conservation. They also facilitated planning and feedback meetings and workshops, conducted interdisciplinary PRAs and other activities, developed data collection forms for farmers and extension staff, and compiled reports and publications. When necessary, contracted experts undertook demonstrations/research in their areas of specialization. Extension staff maintained close interactions with farmers and facilitated farmer activities. They also undertook research and demonstration, data collection and compilation, reviewed planned activities with farmers, assisted in the preparation of training materials for expert farmers to other farmers, and planned and took the lead in farmer exchange visits. As in the objective-setting stage, leaders played an important role in sensitizing farmers and being actively involved in most project activities.

Interdisciplinary working: Challenges and lessons learned

Implementing a project with a multidisciplinary team from diverse backgrounds and experiences is a challenge. Farmers, for example, considered scientists as being strong in formal knowledge with little practical field experience. By contrast, scientists considered farmers as people who found it difficult to comprehend introduced technologies. These salient beliefs and attitudes could only be heard when talking to individuals, but not in the open. Overall, there were challenges to interdisciplinary working related to professionals alone, professionals and farmers, and challenges between farmers themselves.

Professional interdisciplinary challenges

Some professionals found it difficult to cope with changes in approaches and methodology and preferred to use their own standard procedures. The first group of botanists involved in the project were used to undertaking plant species assessment in natural ecosystems, not managed or planted fields. They preferred to use the modified Whittaker methodology (Stohlgren et al, 1995) as an established national approach for plant species assessment. They were reluctant to work with the approach developed in PLEC (Zarin et al, 1999) for the same purpose. Another team of botanists agreed to work with this methodology.

Other researchers were uncomfortable with direct experimentation on farmers' own fields. Most were used to conducting experiments on small plots at research stations and to transferring promising technologies onto farms for farmer evaluation and adoption. Farmers preferred working with bigger plots to be able to assess performance and get immediate benefits. The PLEC project emphasized working with farmers by improving what they were already doing. The research was led by farmers with close facilitation by scientists. The success of the introduced and improved interventions was attributed to farmers taking the lead and becoming teachers of other farmers. It took time for scientists to confidently work directly on farmers' fields.

Researcher–farmer interdisciplinary challenges

Scientists did not believe in experiments conducted by farmers. They did not even know that farmers conduct experiments. Such experiments have significantly contributed to the development of technologies found on farmers' fields. Table 7.1 illustrates a number of farmer-initiated experiments made to address production constraints in the absence of scientific recommendations that farmers preferred.

Table 7.1 *Farmer experimentation to address production constraints*

Experiment	Objective	Duration	Results
Maize breeding of hybrid 622 vs. local *larusa* variety	Improving the *larusa* local variety by using hybrid 622 variety	Two years	Local variety of *matatu* developed with the following qualities: better milling qualities, sweet small cob with many grains, post-harvest pest tolerant, more tolerant to lodging when well spaced
Cauliflower production using different fertilizers (ordinary soil, ash, chicken manure, forest soil, sulphate of ammonia at planting and after second weeding)	Comparing industrial fertilizer and locally available fertilizers for crop production; testing organic production of vegetables	One season	Poorest performance with ordinary soil, premature leaf shedding with ashes, highest yield with chicken manure followed by farmyard manure, others comparable
Maize/beans intercrop spacing	Determine best spacing for optimal production of maize/bean intercrop	Two years	Spacing for sole maize crop of 75 × 30cm and sole bean crop of 45 × 15cm changed to 90 × 30cm maize and 30 × 15cm beans for the intercrop

Plant breeders, for example, did not trust the procedures used by farmers to make their own breeds. In breeding a variety of *matatu* (maize) (see Table 7.1), the farmer used experience gained from on-station research visits. He was successful in developing a variety with better qualities than either the local *larusa* or the improved HB 622 varieties. While the HB 622 hybrid was high yielding, it was not sweet and was susceptible to storage pests. The local *larusa* was sweet but low-yielding with poor milling qualities. The new *matatu* variety was quickly adopted by other farmers, leading to it being included in a project on germplasm collection for further work. Many other ways of farmers pioneering research to address production constraints are discussed by Kaihura (2002).

One challenge to farmers and scientists working together was low acceptance by farmers of technologies recommended by research or extension workers. Most farmers considered scientists' technologies as no better than technologies developed by the farmers themselves, based on their own research, experiences from friends, neighbours and other sources. In such situations, scientists and extension staff had to refrain from influencing farmers' debates and to learn from their models. Conversely, many scientists did not believe that farmers were knowledgeable and capable of making valuable contributions in interdisciplinary teams. The same was true for extension staff who initially expressed superiority to farmers. For example, during discussions while traversing Kiserian village land to assess soil types, qualities and use, one of the elderly farmers was able to identify and characterize soils based on his own indicators. Table 7.2 summarizes the classification of some of these soils using both local and scientific classification indicators. It was realized that some of the farmer's criteria were similar to those used by scientists, and that, while farmers are knowledgeable, their knowledge is not documented. However, some scientists are still not convinced that such farmer-held knowledge can be reported in scientific publications.

Interdisciplinary farmers working with specialized facilitators

In trying to avert production risks, small-scale farmers respond to production uncertainties through diversified farming. Figure 7.1 illustrates a typical complex and dynamic intensification of agriculture in peri-urban Ng'iresi village in the sub-humid zone, characterized by high population pressure with acute land shortage. The farmer in question has different field types for crops, forage, trees, livestock and fish. While land shortage is a factor, she also chooses which crops to grow in response to market demands in Arusha municipality. The field types on the farm include green maize for roasting in town, and vegetables for urban markets. Chicken, hybrid goats, turkeys and rabbits are mostly sold to hotels in town. The farmer also grows tree seedlings for various individuals and organizations working in environmental conservation. Banana/coffee is a typical agroforestry farming system in the area. Guatemala grass (*Tripscam laxum*) and herbaceous legumes on contours are for feeding livestock. Flowers are grown in rotation with a mixed crop of maize and beans. Cattle provide both milk and manure. Some of the slurry from the cattle pen is directed to the fish pond to feed the fish; it also protects the fish from birds by darkening the water. Fish are harvested about twice a year and sold at high price in an area without any reliable source of fish.

Such a complex and diverse system requires a diversity of professionals to advise the farmer, and equally, under such situations, professionals get to learn from farmers. This particular farmer, for example, faced with land scarcity, keeps both layer and broiler chickens under the same roof, feeding them the same feed, until the broilers are ready for market. They are then removed and

Table 7.2 *Comparison of indigenous and scientific soil classifications in Kiserian village, Arumeru*

Soil name		Fertility rating		Indicators	
Local	Scientific	Local	Scientific	Local	Scientific
Engulukoni nanana/ nabuyavi	Eutric Andosol	Very high	Very high	Very fertile dark soil with good moisture holding capacity, used for raising tree seedlings in nurseries	15–20cm deep decomposed surface litter, very dark brown to very dark greyish brown clay to silt clay
Engulukoni narok	Haplic Andosol	Low	Moderate	Surface truncated soils due to erosion, need inputs to maintain good yields	Deep dark brown clays with weak structure, with strongly weathered basaltic rock below, on summits and steep slopes
Engulukoni nador	Haplic Luvisol	Low	Low	Poor fertility soils that need inputs like organic matter or fallowing to realize good yields	Weakly structured dark reddish brown clays on upper and middle slopes
Engulukoni nador	Haplic Nitisol	Moderate	Moderate	Soils prone to water and wind erosion with reduced fertility	Deep red clays with shiny faces
Engulukon sero	Vertic Luvisol	Moderate	Moderate	Clays that need controlled erosion and inputs for good production. Need timely tillage operations	Intergrade between reddish brown clays and heavy deep black clays, cracking when dry with salty subsurface in lower slopes

the feed is changed to mash for the remaining layers. While experts were never trained to combine keeping different types of chickens in one room, they accept that this is a sound practice and are now trying it with farmers in other areas facing land scarcity. Through different constraints and possible production risks, this farmer has developed skills and knowledge to manage different enterprises together on a small piece of land. She is more skilled than experts and one single extensionist is not competent enough to bring about desirable improvement to the farm; an interdisciplinary team would have to be involved.

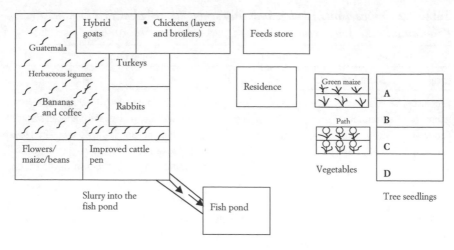

Figure 7.1 Diversified production in land-scarce peri-urban Ng'iresi village

Farmer–farmer interaction challenges

When farmers are selected for participation in externally funded rural development projects by extension staff or village leaders, resource-poor farmers are usually ignored. However, such farmers are not necessarily handicapped in knowledge and experience. They have something to share. The well-off farmers – who are repeatedly selected by village leaders to participate in introduced projects – did not initially believe they could learn from the poor, and the same was true for several scientists and extension staff. The poor were rarely given opportunities to provide their views at farmer-to-farmer training sites. They did not seem to have anything to offer when visited. PLEC took the challenge of bringing them on board. Most had rich biodiversity on their farms as one strategy of risk aversion. They also had lessons for wealthier people during discussions. Several had opportunities to address workshops and other people through the mass media. The full involvement of poor people, and the benefits they enjoyed by involving themselves further, made the project popular and attractive to farmers beyond the boundaries of the project area.

One research issue that led to real challenges for some farmer families was open accounting of income and expenditure at the household level. Based on the annual calendar of activities, project researchers proposed to establish at which times of the year households had a surplus and when they were in deficit, in order to plan for reinvestment in agriculture. Income and expenditure forms were introduced to several families to record all incoming and outgoing money every day. The exercise indicated that most men spent the household income on drinking – something women did not like. It became a source of conflict to some families and a threat to continued participation in the project. Some did end their involvement in the project, but others improved their income/expenditure balance and managed to reinvest.

PRA conducted in 1996 (Kaihura, 1996) revealed that, due to poor farming on sloping farms in the uplands, the most fertile soil was being lost to the lowlands, and farmers there harvested good yields from this soil at the expense of the upland farmers. Upland farmers were ashamed and planned to dissociate themselves from lowland farmers until the erosion was controlled. This was a threat to project activities involving farmer exchange visits and farmer-to-farmer training programmes. Because each group had relatives in both the lowlands and the uplands, and upland farmers had temporary plots in the lowlands and exported nutrients from the lowlands to the uplands through grain and stover (crop residues), it was possible for PLEC to normalize the situation.

Main project outcomes at the local level

Improved on-farm biodiversity

Improved on-farm biodiversity resulted from sensitization and demonstrations of intercropping, mixed cropping, conservation farming, agroforestry and improved pasture and woodlot management. For these technologies, new materials – both improved and indigenous – were introduced.

Crops that had disappeared due to a lack of market, such as sweet potatoes and cassava, were reintroduced, linked to strategies of adding value through processing. New livestock breeds (goats, pigs, local chickens) were introduced to improve local breeds.

Improved agricultural intensification

The common practices of leaving many banana stems on one stool and random mixed cropping in land-scarce areas changed. There were improvements resulting from recommended crop spacing, proper combinations and sequences of crops, and associated management of each crop in the crop mix. When there were many bananas on one stool, these were previously kept for cutting and feeding to stall-fed cattle. Stall feeding of banana stems was condemned during a workshop by farmers who buy milk from these farmers; this helped improve banana production.

Increased yields and reduced soil erosion

The improved management and use of manure and compost, conservation farming on steep slopes, and improved crop livestock integration (for example, adding bean stover into kraals as bedding before application in the field) increased the weight of banana bunches from 30 to 50kg. In semi-arid

Kiserian, yields of maize grain increased from 250 to 1500kg/ha through traditional means of erosion control using stonelines. Conservation farming on steep slopes improved surface cover, fodder yields along contour lines, and the diversity of fodder on contours. Besides erosion control, the diversification of contours with fodder materials improved the quality of feeds and milk yields from stall feeding. Income and nutrition improved at the household level, as did on-farm biodiversity.

Improved nutritional value and livelihoods of smallholder farmers

The successful management of introduced income-generating and biodiversity-enhancing activities changed ordinary farmers' diets and livelihoods. Although many farmers hesitate to slaughter chicken for meat, they willingly sell eggs and chicken to earn money. They also give some eggs to children as food. Through egg sales, the very poor paid for children's school fees and uniforms and met basic household requirements such as salt, paraffin and clothing.

Renewing degraded pastures and forests increased honey production

Through farmer groups established by farmers themselves, tree seedlings were raised in nurseries and planted in degraded pastures and open homesteads. Along with raising tree seedlings in nurseries, knowledge of how to grow indigenous trees of economic importance was passed to youngsters from the elderly.

Reforested and planted forests were protected and beehives were introduced in the forests, mostly on steep slopes and water catchments. As a result, grazing, firewood collection and the burning of forest bushes were greatly reduced. In addition, honey was and is still being collected for home use and selling. Several farmers also use honey for medicinal purposes.

Effective and improved land utilization

Land facets such as farm boundaries were better planned for production by combining trees, fodder and crops along a single boundary line. The emphasis was on utilizing land to produce fodder to feed livestock by planting *Sesbania*, and providing both construction timber by planting *Grevillea* and food by planting yams that creep on the *Grevillea*. Improved food production also resulted from using man-made soils with high organic matter. Crops with edible roots and shoots are planted in these man-made soils and can be harvested when needed. Another innovation was effective sequential cropping that ensures continuous harvesting and space optimization, including both annuals and perennials.

Establishment of by-laws

The project sensitized farmers to the need for those who exploit conserved areas to share benefits from conserved biodiversity. By-laws persuading the Tanzania Electrical Supply Company (TANESCO) and timber companies who use water from village-conserved forest catchments for hydropower, and timber and furniture, respectively, mean that payments are now made to villages. A primary school has been built in one of the villages, and roads destroyed while transporting logs have been repaired.

By-laws on farmers' intellectual property rights on the conservation and use of natural and managed biodiversity were proposed. Other by-laws considered the transfer of ownership and responsibility of the conservation of volcanic hills from local government to neighbouring villages, and recommendations to improve quality of livestock stall feeding and marketing of milk.

Women's empowerment

Community and farmer groups decided to create opportunities for women to freely participate in community development activities and, where possible, to be given leadership responsibilities. Women had to volunteer to take leadership responsibilities. Women are now heading farmer groups and are among the lead persons in community development activities. Most groups led by women perform better than groups led by men.

Inclusion of environmental studies in primary school curriculum

Some primary schools involved their pupils in environmental conservation through raising tree seedlings, and planting and managing them at school and at home. The look of the school compounds has changed as a result of established woodlots, and trees have also been introduced at each child's home. The objective was to gradually convince the Ministry of Education to include practice, rather than theory, of environmental conservation in the school curriculum.

Establishment of good working relations and farmer networks

The initial scepticism of farmers regarding knowledge and recommendations from scientists – as well as scientists' perceptions of the knowledge and technologies of farmers – changed for the better. Farmers and scientists worked as equal partners, sharing all the secrets behind successes and failures in production in open discussions. Farmer-to-farmer relations also improved, and the exchange of information and knowledge through individual and extension-facilitated field exchange and training visits became formalized.

Identified expert farmers were enabled in terms of communication skills and materials to teach other farmers. The number of trainer expert farmers increased from five at the start of the project to ten at the end. The awareness of farmers regarding sources of information on improved technologies and on-going scientific research was improved.

Main project outcomes at other levels

Mobilization of other farmers

Regional and district leaders identified established farmer groups as key instruments to sensitize farmers elsewhere in aspects of biodiversity and environmental conservation by means of exhibitions of PLEC farmers' products, organized drama and singing, and supporting expert farmers to train other farmers. Farmer groups from outside the project area have been mobilized and brought to PLEC sites, and PLEC farmers have been invited to visit other farmer groups.

Testing of participatory methodology of PLEC

The last annual planning workshop made a resolution that the PLEC approach of 'participatory technology development and dissemination' (PTDD) be further tested along with other participatory methods in order to develop a national cost-effective, rapid-impact approach.

Interactive opportunities with leaders

Through the project, farmers were able to interact with representatives from input supply companies, institutions and policy-makers in order to discuss production constraints associated with the farmers' own situations and available technologies.

Continuation of PLEC work after the project

Due to the relevance of PLEC to rural livelihoods, farmers requested a second phase of the project. They established PLEC management teams to continue PLEC work – and this continues until today. In addition, project staff at the local level have, since PLEC, been identified to lead the mobilization and facilitation of farmer groups in the district. Scientists involved in PLEC have been invited to different workshops to share project experiences with other projects. Others are currently working as consultants in ongoing and newly developed projects to extend good PLEC experiences to the new projects.

Development and implementation of recommendations

Local and national stakeholders and project implementers developed recommendations during field days and feedback and planning workshops. Some of the recommendations were of local relevance, while others were to be followed up at national level. From 1998 to 2002, annual project workshops and six-monthly feedback meetings were conducted at the project sites with PLEC farmers and other stakeholders in the district. After every workshop and feed-back meeting, specific observations and recommendations were developed. Some of those that needed implementation at village and district levels were addressed immediately by the district. Those that needed to be incorporated into national plans and/or budgets were presented during the technical and policy recommendations workshop, at the end of the project, to national-level policy- and decision-makers for consideration (Kaihura and Stocking, 2003). Some of these are now being implemented.

One example of the development and implementation of technical recom-mendations is with regard to climatic change. Studies on long-term spatial and temporal changes in rainfall in Arumeru (Kingamkono and Kaihura, 2003) indicated a general decrease in total rainfall and the length of growing season, as well as changes in the onset and cessation of rains. A national strategy was proposed to create awareness among farmers and land users of the revealed trends and develop coping strategies.

Another recommendation derived from the observation that commercial companies from outside the area exploited the forests and left no benefits to the village and local people who are the key conservationists of observed bio-diversity and rehabilitators of degraded lands. It was recommended that the farmers' conservation efforts should be recognized and that commercial companies like TANESCO and the National Urban Water Authority (NUWA) should return some of the benefits to rural communities as incentives and to use for rural community development. Thus, support for farmers, rather than outside commercial interests that return nothing to the local community, would bring substantial national benefits both in protecting ecosystem service functions and enhancing local rural development.

In relation to observed unsustainable land management practices, it was recommended that current practices of clearing natural forests and planting commercial ones should be recognized as unsustainable and unable to support national and local economies. Most afforestation in Arumeru is directed to humid and sub-humid areas and concentrates on clearing natural forests in order to replace them with planted exotics. This reduces biodiversity and agro-diversity benefits in smallholder farms. Future plantations should avoid monocultures and include indigenous trees. Future afforestation should be directed to areas of sparse vegetation and, in particular, deliberate large-scale afforestation of semi-arid lowland environments, especially with indigenous trees, should also be initiated by responsible institutions. Meanwhile, the potential of indigenous tree species and pastures should be emphasized, partic-ularly with respect to biodiversity conservation, fertility restoration, erosion control and their use in agroforestry systems.

A number of policy recommendations were made in relation to the implementation of several articles of the Convention on Biological Diversity (CBD). Despite several government initiatives regarding environmental management and biodiversity conservation, the legislative process to introduce several of the CBD articles has not yet been concluded. Farmers expressed concern that they were the most disadvantaged, as their indigenous knowledge, practices and genetic resources were being exploited without legal controls. Technologies developed by local farmers to use indigenous tree species to control pests and diseases in crops and livestock were a case in point: outsiders seek details of species and practices without promising any benefits in return to farmers. There are no provisions for access or transfer of knowledge to outside parties with commensurate recompense to the local custodians of biodiversity. Members and stakeholders of PLEC and its activities recognize that agrodiversity encompasses more than the monitoring and conservation of biodiversity; it takes into account the people who manage the diversity, the management systems, the biophysical environment in which plants grow, and the organizational aspects of conservation and management. Indeed, it supports human needs and development. The recommendations with regard to the implementation of the CBD covered; first, improvement of local knowledge of the CBD; second, creation of a national database on biodiversity and sharing information between relevant ministries; third, establishment of by-laws to protect small-scale farmers; fourth, incorporating farmer-desired qualities in plant breeding; fifth, government support and facilitation of knowledge exchange between farmers; and sixth, strengthening cross-sectoral groups.

Other outcomes

Other outcomes included:

- the increased capacity of research scientists to undertake research on biodiversity-related issues;
- improved understanding and skills for researchers and extension staff to work with farmers;
- contributions to the national database on agrodiversity components (agrobiodiversity, biophysical diversity, management diversity and organizational diversity) in representative biodiversity hotspots;
- contributions to methodologies for on-farm biodiversity assessment and analysis;
- widely distributed publications: newsletters, books, journal articles, training materials and leaflets, as discussed below.

The dissemination of project results and outcomes

The various groups of stakeholders were involved in different ways in disseminating the results and outcomes of the project. For farmers, this included

informal and formal farmer-to-farmer visits and discussions on the performance of different technologies; swapping of materials like crop varieties,
medicinal herbs, and hedgerow and conservation structures; exchange of information and experiences between farmer groups; and exchange of information
between project and non-project farmers during exchange visits. Expert farmers participated in national agricultural shows, agricultural exhibitions at
community and district levels, and national and international conferences and
workshops, making it possible for them to interact and display or demonstrate
some of the project outcomes.

Both political and religious leaders at village, community and district levels
were involved in sensitizing farmers to be actively involved in the project,
spreading promising technologies to their areas through meetings, political and
religious speeches, and so on. Representatives of the Ministry of Agriculture
and Food Security, Department of Environment of the Vice President's Office
and the national Environmental Management Council, for example, participated in annual and national workshops. Learning from the project's findings
and recommendations, they could advocate the wider adoption of the technologies developed in the project. Some also participated in local and policy
recommendations workshops. Researchers and extension staff produced and
distributed leaflets about technologies developed in the project, and facilitated
the participation of farmers in national agricultural shows and other local and
district-level exhibitions.

The media were also important. Both radio and TV stations broadcast
workshops, field days and on-farm farmer-to-farmer training programmes. For
instance, the Ministry of Agriculture and Food Security's monthly TV and radio
programme 'Ukulima wa Kisasa' (Improved Farming) promoted dissemination
of project outputs. Newspapers reported project events and recommendations.
Internationally, the project led to articles in the PLEC project newsletter, *PLEC
News and Views*, of which 20 issues were released over the four years of project
implementation. Issue 13 was a special issue on methodology; other key project
publications included Zarin et al (2000) and Kaihura et al (2002a, b); there
were also books, leaflets and training manuals such as Brookfield et al (2002,
2003), Gyasi et al (2004), Kaihura and Stocking (2003) and Stocking and
Murnaghan (2001) (see the PLEC website www.unu.edu/env/plec/ and
http://rspas.anu.edu.au/anthropology/plec.html).

Conclusions and recommendations

Farmers are multidisciplinary in nature. They manage a diversity of farm activities ranging from soil, crop and livestock management to woodlot and apiary
management, and fish keeping. Interdisciplinary research teams are the most
appropriate in projects that aim to work towards the improvement of farmers'
production systems and livelihoods. However, deeper understanding of social
and cultural behaviour in this PLEC project was limited by a shortage of

anthropologists and socio-economists.

Farmers employ both technologies developed by experts and others from their own innovations. They also conduct their own research. Through the PLEC project, they demonstrated that they are knowledgeable and provide resourceful inputs into interdisciplinary teams addressing farmers' problems. They also have experiences and practices from which professionals can learn. Some of them are experts able to train other farmers. Their knowledge and practices are, however, not documented. There is a need to make deliberate efforts to document and capture this treasure of knowledge and ensure their full participation in research in order to realize sustainable results.

Through repeated interactions between scientists, farmers and other stake-holders, great impact was achieved in terms of improved resources management, food security and livelihood improvement. It was also possible for farmers to continue project work after the project. By being able to bring about change in peoples' attitudes it was easy to realize the desired impacts within the lifetime of the project.

Funding organizations usually claim that repeated field visits are expensive. However, this project showed that rapid results and impacts resulted from continued on-farm interactions of researchers and farmers. With farmers as leaders, technologies are rapidly disseminated beyond project boundaries. Farmer-to-farmer working relations and interactions also improved. The benefits of continued scientist–farmer interactions are far greater than the costs involved and should not hinder the funding of such projects.

The implementation of research is greatly improved by involving all relevant stakeholders, particularly those in the project area. In this project, such broad involvement improved working relationships between projects, reduced confusion for farmers addressed by several projects, and encouraged majority participation – especially when religious leaders and policy-makers participated.

The poor and women are key to project implementation. It is unfortunate that the poor have been largely omitted in many projects, while women are greatly under-recognized by society. Since the very poor are those who most degrade the environment, and women are the principal actors in livelihood improvement, they should be deliberately focused on in order to realize sustainable rural development.

Finally, changes in land-use and cropping systems in the uplands were linked to some extent to changes in the lowlands and vice versa. Understanding these dynamics and interactions was important for the development of recommendations for rational resources management. This can best achieved if projects are implemented at the landscape scale.

Acknowledgements

The farmers and scientists are grateful to the Global Environmental Facility for funding the project and the United Nations University for implementing the project. Many thanks to the Banff Centre and the UNU for supporting my participation in the IRMMA conference.

References

Brookfield, H., Padoch, C., Parsons, H. and Stocking, M. (eds) (2002) *Cultivating Biodiversity: Understanding, Analysing and Using Agricultural Diversity*, London, ITDG

Brookfield, H., Parsons, H. and Brookfield, M. (eds) (2003) *Agrodiversity: Learning from Farmers Across the World*, Tokyo, United Nations University Press

Bureau of Statistics (1998) *Population Census*, Dar es Salaam, Bureau of Statistics

Gyasi, E. A., Gordana, K., Blay, E. T. and Blay, O. W. (eds) (2004) *Managing Agrodiversity the Traditional Way: Lessons from West Africa in Sustainable Use of Biodiversity and Related Natural Resources*, Tokyo, United Nations University Press

Kaihura, F. B. S. (1996) *Farming Systems Response, Biodiversity and Adaptation to Conservation in Tanzania*, East Africa PLEC Tanzania pilot phase report, Tanga, Tanzania, Agricultural Research Institute

Kaihura, F. B. S. (1999) *Potentials and Strategies for Alleviating Soil Fertility Decline in Kagera Region Using Integrated Plant Nutrient Management Systems*, presented at the workshop on 'Integrated Pest and Plant Nutrient Management in Kagera', Kagera, Tanzania, Kagera Environmental Management Project (KAEMP)

Kaihura, F. B. S. (2002) 'Working with expert farmers is not simple: The case of PLEC Tanzania', in Brookfield, H., Padoch, C., Parsons, H. and Stocking, M. (eds) *Cultivating Biodiversity: Understanding, Analysing and Using Agricultural Diversity*, Tokyo, United Nations University Press

Kaihura, F., Kaitaba, E., Kahembe, E. and Ngilorit, C. (2002a) 'Tanzania', in Brookfield, H., Parsons, H. and Brookfield, M. (eds) *Agrodiversity: Learning from Farmers Across the World*, Tokyo, United Nations University Press

Kaihura, F. B. S., Ndondi, P. and Kemikimba, E. (2002b) 'Agrodiversity assessment and analysis in diverse and dynamic small scale farms in Arumeru, Arusha, Tanzania', in Brookfield, H., Padoch, C., Parsons, H. and Stocking, M. (eds) *Cultivating Biodiversity: Understanding, Analysing and Using Agricultural Diversity*, London, ITDG

Kaihura, F. and Stocking, M. (eds) (2003) *Agricultural Biodiversity in Smallholder Farms of East Africa*, Tokyo, United Nations University Press

Kaitaba, E. G., Kaihura, F. B. S., Kiwambo, B. and Mowo, J. G. (1999) 'Profile descriptions of soils of Arumeru', in Kaihura, F. B. S. (ed.) *Farming Systems Response, Biodiversity and Adaptation to Conservation in Tanzania: The Case of Arumeru District, Arusha*, EAPLEC Tanzania Pilot Phase Report

Kingamkono, R. M. L. and Kaihura, F. B. S. (2003) 'Spacial and temporal characteristics of rainfall in Arumeru district, Arusha region, Tanzania', in Kaihura, F. and Stocking, M. A. (eds) *Agricultural Biodiversity in Smallholder Farms of East Africa*, Tokyo, United Nations University Press

Morss, E. R. (1980) *Cross-cutting Issues Emerging from the Arusha Regional Planning Exercise*, Paper presented at the AP/VDP (Annual Plan/Village to District Development Planning)

Stocking, M. A. (2003) 'Tropical soils and food security: The next 50 years', *Science*, vol 302, pp1356–1359

Stocking, M. A. and Murnagha, N. (eds) (2001) *Handbook for the Field Assessment of Land Degradation*, London, Earthscan

Stocking, M. A., Kaihura, F. and Liang, L. (2003) 'Agricultural biodiversity in East Africa: Introduction and acknowledgements', in Kaihura, F. and Stocking, M. A. (eds) *Agricultural Biodiversity in Smallholder Farms of East Africa*, Tokyo, United Nations University Press

Stohlgren, T. T., Falker, M. B. and Schell, I. D. (1995) 'A modified Whittaker nested vegetation sampling method', *Vegetatio*, vol 117, pp113–121

Zarin, D. J., Guo, H. and Enu-Kwesi, L. (1999) 'Methods for the assessment of plant species diversity in complex agricultural landscapes: Guidelines for data collection and analysis from the PLEC Biodiversity Advisory Group (PLEC-BAG)', *PLEC News and Views*, vol 13, pp3–16

8

The Mountain Mistra Programme: Options for Managing Land Use in the Swedish Mountains

Tomas Willebrand

Location, natural environment, human settlements and activities

The Mountain Mistra research programme focuses on the mountains of north-western Sweden, particularly the 15 mountain communes that represent 41 per cent (165,000km²) of Sweden (see Figure 8.1). Most of this land is covered with boreal coniferous forest; about 40,000km² is alpine or tundra. The region contains large wilderness areas and conservation values of international importance; most of Sweden's national parks are situated here. In addition to the 6500km² protected in national parks, at least twice this area has some form of nature conservation protection. This region is the home of all the remaining large carnivores, especially the wolverine, which is primarily found in the borderland between the coniferous forest and alpine mountains.

All Swedish communes have an elected council that has primary responsibility for planning the use of land and water, under the supervision of the national government. The communes must, however, consider national interests when drawing up their plans. There are also powerful regional county administrative boards, which are the regional branch of the government but also assure a regional influence on national decisions. The 15 mountain communes are divided between four regional administrative boards. Many decisions on natural resource management – for example, with regard to reindeer herding, hunting, fishing and forestry policies – are implemented at this regional level, where there is a blend of state agencies, directly elected institutions and municipal associations. A large proportion of the mountain range (60,000km²) is state-owned land, directly managed by the regional administrative boards. Most of the forest previously owned by the state Swedish Forest Service is now managed by the state-owned company Sveaskog.

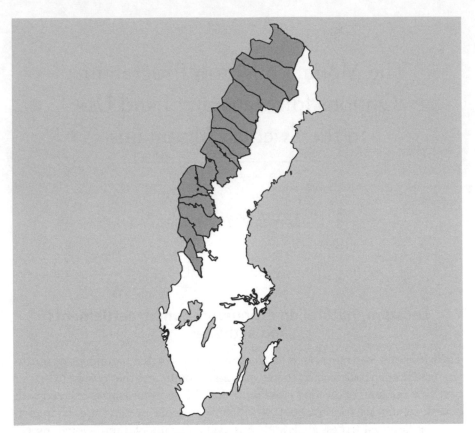

Figure 8.1 The 15 mountain communes with elected councils

The Swedish mountain region is characterized by a sparse population, comprising less than 2 per cent of the Swedish population (150,000 residents live in an area of 145,000km²). The population in this region is declining due to a decline in death and birth rates, and emigration to the larger cities in southeast continues, though at a reduced rate. Forestry used to be the major source of employment in this region, but today less than 10 per cent are employed in this sector. Tourism is believed to have great potential benefits, but has not yet reached the employment rates of the forestry sector. The local municipalities are the most important employers, either directly in administration or indirectly through schools, health care and so on. This transition is creating shifts in cultural outlook, adjustments in economic activities, and reconfigurations of social networks and political alliances. Though few of the inhabitants in these areas own much land, they still have strong ties to it. The principle of common access to the land, and the possibilities of using its resources, such as berries, wood fuel, fish and wildlife, for modest fees are strong. One could even say that the people and the land constitute the social system in this region.

One of the oldest living cultures in the region is that of the Sami people who have been surviving in these barren landscapes for thousands of years. They traditionally lived by hunting and fishing, and by reindeer herding since the late 17th century. There are about 17,000–20,000 Swedish citizens that consider themselves as Sami. The Sami Parliament was founded in 1993 as a government authority with 31 members, elected in a general election by Sami entitled to vote. The Sami Parliament in Sweden has a double role, being both a popularly elected body and a government authority. The Sami have especially strong links to the land due to their culture and long tradition in the area.

Only Sami have the right to herd reindeer in Sweden, and only those who have a membership in a Sami 'village' can exercise this right. The Sami are dependent on the possibility of herding reindeer over large areas, from the high mountains in the west during summer to lichen-rich old conifer forests in the east during winter. A Sami village is both an administrative management unit and an economic corporation working for the benefits of its members. There are 51 Sami villages and about 2500 reindeer herders in Sweden. As there is a limit to how many reindeer can be herded in each village, there are currently about 240,000 reindeer in Sweden. Like all local residents, the Sami also depend on subsistence fishing and hunting.

Harvesting of fish and wildlife was historically the most important land use in the mountain region, but subsequently reindeer husbandry and agriculture became more important. The mining industry also became important for the inhabitants of the area at a relatively early stage during the late 19th century. Forestry and hydroelectric development became especially important in the latter part of the 20th century. However, over the last 20 years the importance of forestry and dam construction has declined, while tourism has increased in importance. All the development has been based on natural resources over vast areas where resources are low in density, and where production and resilience are likewise low. Pressures on these resources are increasing, together with an awareness of the necessity of conserving biodiversity. This results in increased competition for, and conflicts over, resources. High levels of unemployment, a declining population and the increasing needs of modern man lead to stress, desperation and conflicts.

A distinguishing mark of this region is that both past and current land-use patterns have been very dynamic, and this will probably continue. Another characteristic is that groups from outside the region often have strong opinions about the management of its resources, which is unusual when compared to other regions. 'Outsiders' can include visitors or even the government, which owns large areas in the region. The values of wider society may thus override the opinions and values of local people, creating a sense of powerlessness in the local communities. In particular, there is an inclination for people not living in the region to focus on conservation, whereas local people tend to focus on the use of natural resources, and this creates a tension underlying many conflicts (see Figure 8.2).

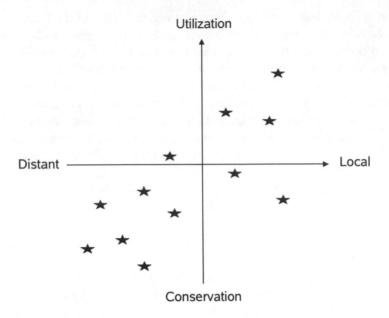

Figure 8.2 Schematic representation of how land-use values
appear to change with distance to resource

Initiation, planning and evolution of the programme

The Foundation for Strategic Environmental Research, Mistra, created in
1994, supports strategic environmental research with a long-term perspective,
aiming to solve major environmental problems. Most of Mistra's funding is
focused on broad-based interdisciplinary programmes. Mistra views these as a
meeting place between two worlds: the research community and the everyday
world driven by the need to solve environmental problems. Here, environmentally sound products, services and production processes must earn money
in a market, and conflicting interests need to coexist. Each Mistra programme
should bridge the boundaries between research and practical implementation.
All are headed by an executive committee, and Mistra keeps track of each
programme during its entire lifetime; its secretariat meets representatives of the
programmes several times a year.

Vision and goal of the Mountain Mistra programme

The vision of the Mountain Mistra programme is that future management of
natural resources in the mountain region, and any external disturbance of these
systems, may best be understood from the perspectives of the most relevant
disciplines, including ecology, economics, sociology, political science and
historic–cultural studies. Policy decisions and changes in land use should have

a common conceptual framework in which the gains and losses for different groups can be quantified and discussed. Conflicts between different groups, as well as the character of the underlying factors causing the conflicts, are thus better understood, and a common view of the possibilities and limitations of different forms of land use in the Swedish mountain region becomes considerably strengthened. These changes should ensure the long-term development of the region and reduce the risk of crossing unexpected environmental thresholds, thus promoting an environmentally sound society and the sustainable use of natural resources. The programme goal is to develop scientifically based strategies for the management and long-term development of the mountain region's resources. This goal includes a detailed architecture for knowledge exchange, and emphasizes interactive projects with users to ensure that the programme's findings are dispersed as broadly as possible. The framework incorporates not only knowledge, concepts, tools and examples that cut across scientific disciplines, but also input from stakeholders.

Outlines of future priorities in land use are used as focal points for the programme team's deliberations and continued interactions with relevant stakeholder groups. These projections serve to crystallize thinking among researchers from different backgrounds and provide a natural avenue for stakeholder involvement. An adaptive approach to scenario development, in which stakeholders actively participate together with researchers, has enabled the programme team to dynamically refine our modelling efforts and effectively communicate key results to interested parties. The successful implementation of this approach requires that the projections are developed with advanced technology and clearly communicated. The main idea is that, by deepening our understanding of the different perceptions among different interests and users of the existing and/or potential conflicts over resources in the mountain region, the programme may contribute to society's ability to solve these problems.

Planning and phase one

The programme was originally planned for a ten-year period and eventually had three phases: the initial phase (1998–1999), phase one (2000–2002) and phase two (2003–2006). Mistra financing had the advantage of making possible an extended research period where complex dynamic processes, such as those in the mountain region, could be incorporated with greater ease than in more traditionally funded projects. Mistra has since revised its policy and most programmes now obtain, at most, funding for planning and then two three-year phases.

In 1995, an initial proposal was submitted by a group of researchers in ecology and forest economics to Mistra, which then allocated a planning grant to the programme. The initial proposal was followed by many in-depth discussions with stakeholders during 1996 that resulted in a new programme proposal in 1997. Mistra granted SEK8 million (US$1.16 million) to the initial phase (1998–1999). While a few research projects started in this phase, most of the work focused on communication and integration strategies, both within the programme and with the stakeholders. A proposal for phase one was submitted in 1999, and Mistra allocated SEK40 million (US$5.8 million) for this first phase (2000–2002). Following an evaluation and reorganization of the programme, it was allocated a further SEK36 million (US$5.2 million) for an additional three years. The amounts included an administrative overhead of about 40 per cent.

Phase one of the programme was divided into seven sub-programmes based on a number of disciplinary focuses, such as forest economics, reindeer management, biodiversity, tourism, and fish and wildlife ecology. Work focused on data collection and analysis, with researchers in the programme producing more than 100 articles and reports. 21 of the most relevant publications, selected for the review process, are included in the references to this chapter (those publications dated from 2000 to 2003). Much energy was also devoted to finding additional complementary funding from traditional research councils, which turned out to be difficult to coordinate towards a common goal and to synchronize in time. This first phase can, to a high degree, be characterized by a spirit of scientific optimism in which the researchers involved set out to solve conflicts in natural resource management by producing good science. Science would improve the conditions for many of the people living in the mountain region without making any other individual worse off, a classic Pareto improvement. Resources put into synthesis, integration and outreach were limited in phase one, but this was not due to a clear decision by the director and board of the programme. All could see the importance of these activities to form a common internal and external framework for the programme, but there was a general lack of experience and good examples. At the end of phase one, it was evident that the programme needed a specialized communicator to establish two-way communication with the stakeholders.

Evaluation and metamorphosis

The programme went through two extensive evaluations, one scientific and one that emphasized its value to stakeholders. The discussions and changes resulting from the evaluation process were intensive and dynamic. It was now clear that resources had been cut too thin and divided between too many tasks, with the programme becoming disparate rather then coordinated. The vision and aim of the programme was still correct, but the overall framework for interaction among researchers and exchange with stakeholders had to be revised. The programme needed a structure that could benefit from all the research and experience gained in phase one. Thus, the proportion of ecological researchers was reduced and political science was included as a new major discipline, along with economics.

The beginning of the restructuring was a two-day workshop involving both the researchers and the newly established network of stakeholders. There were 30 participants, with an equal representation of researchers and users. An important task was to identify the most important challenges for the future management of natural resources in the region. What was needed to increase the possibility of seeing alternative options? What would make the resource management options into clear outcomes shared by all, and that could be environmentally robust and politically deliverable?

Based on these discussions, interactive work with stakeholder representatives from the user network, and on phase one research, four areas of research focus (focal points) were identified. Each focal point contained researchers from at least three different disciplines. One researcher was appointed to be responsible for coordinating the work within each focal point. There are no individual projects within the focal points and the programme stresses that it is the group as a whole that has a number of research tasks to execute, not the individual researchers (who are contracted for a specific amount of time in the programme). The development of the focal points is evaluated on a regular basis by the interdisciplinary executive group, comprised of three professors and the programme director, which can make recommendations to the board if needed.

The focal points are:

- *Reindeer industry under pressure.* This focal point deals primarily with effects of competing forms of land use on the availability of and access to grazing land. Reindeer husbandry is strongly affected by other types of land use, especially in forested areas where different landowners expect a profit from forestry and rarely see any benefits to having reindeer on their land. Reindeer are sometimes seen as a potential source of forest damage, and the Sami are often perceived as having weaker rights. Most groups agree that the amount of winter food available to the reindeer is decreasing; this constitutes a threat to the reindeer in the two northernmost counties. The situation is somewhat different in the southern part of the reindeer husbandry area, where the winter areas are threatened by disputed reindeer management rights. Even a small increase or decrease in forestry activities can have major repercussions on the overall carrying capacity for reindeer in a herding district. Thus, the relationship between forestry and reindeer is an important area where different mutual strategies and possible future changes in policies need to be explored (Sandström et al, 2006). Moreover, reindeer are also dependent on the summer ranges, which largely determine the productivity of the animals. Even though the herding districts have better possibilities to manage their grazing areas in the summer areas, threats exist in the form of disturbances from tourism, hunting and fishing. There are also claims that reindeer are too concentrated in some areas of the summer range and therefore may damage the vegetation and increase the risk of erosion. An overview of the complex patterns of different processes affecting the reindeer herding is shown in Figure 8.3.

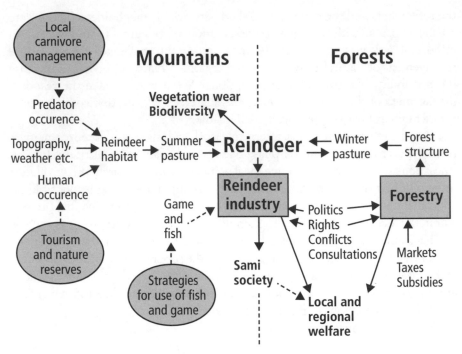

Figure 8.3 Reindeer industry under pressure

Source: Ö. Danell

- *Strategies for the use of fish and game.* This focal point deals specifically with strategies of managing fish and game populations through various rights of use. What are the possible strategies for managing the harvest of fish and game, with emphasis on state-owned land above the so-called 'cultivation border'? Increased access to small game and fish in the local area above the cultivation border is seen as a major problem, together with the question of sharing the quota for moose between Sami and non-Sami populations. There is a need to evaluate the consequences of different management strategies of these resources (Eriksson et al, 2006). The harvest potential can be measured in terms of biomass or, more usefully, as potential user-days that could be utilized by different interest groups. This potentially could aid the Sami community in developing the commercial base to support reindeer herding and in preserving their culture, or it could be given to local enterprises to develop private recreational tourism. Alternatively, it could be claimed as a national resource used to maximize the recreational (welfare) value for the nation as a whole. These decisions on the 'best' strategy could be taken by the central administration or handed down to a more local level, depending on the emphasis placed on the possibilities of and limitations on resource utilization. However, it is important that people at the local level have access to knowledge and resources so as to co-manage the resource with the national level. This focus includes looking at problems of securing a sustainable use of these resources, and of identifying winners and losers in different management regimes.

- *Decentralized management of large carnivores.* This focal point addresses the social and economic values of large carnivores in the mountain region. We are especially concerned with the question of what will happen once the goal set by the Swedish Parliament regarding the populations of carnivores is reached, as the large carnivores and their relation to central/local inhabitants is likely to become a very critical and perhaps a contentious issue (Ericsson et al, 2006). The goal is not framed in terms of any biological equilibrium; keeping the large carnivore populations within a restricted range will require active management and, probably, strong local co-management. Increased local management will need increased knowledge and resources but may find itself undercut by negative and hostile attitudes emanating from urban–rural tension. The central question is: what are the requirements for effective management strategies of large carnivores? Economic and social values are important issues, but the effects on reindeer husbandry, especially disturbances and losses of livestock, also need to be examined.

- *Tourism development and protected areas.* This focal point considers the analysis of policies and plans for the development of tourism as well as the management of protected areas. What are the possibilities of tourism development as a source for economic and social welfare in the mountain region's communities, and how may this affect other user groups and impact natural resources in protected areas? New forms of tourism, increased numbers of visitors, and concentration of tourists in certain areas may create both positive and negative impacts. Positive impacts may include opportunities for socio-economic development through the creation of new forms of tourism related to, for example, hunting and fishing; but there may also be new forms of non-consumptive tourism in protected areas (Zachrisson et al, 2006). Negative impacts could be detrimental environmental effects on the landscape, aesthetic values, biological diversity or the degradation of sensitive vegetation. Furthermore, greater numbers of tourists may also increase congestion, reduce local influence on land use, and create new conflicts between users – including between different types of visitors and within visitor groups. The current emphasis on developing certain strong tourist centres may also create economic difficulties for smaller-sized tourist areas in the mountain region. As for the protected areas, there are trends in nature tourism that may have implications on their future management. Diverging views on nature among different user groups, the right of common access to land versus nature conservation values or the private sale of rights to hunting and fishing, along with the demand for increased local management of protected areas, could all lead to new types of conflicts and consequences.

There was also a need to develop both conceptual models and theoretical support within the research areas. We chose to develop aspects that correspond to the three different theoretical perspectives of research disciplines: ecology, economics and political science. These analytical frameworks should provide theoretical perspectives in all four focal points presented above. Our analytical framework consists of a three-pronged approach within natural resource and welfare economics, natural resource ecology, and democracy and public administration of natural resources. These three approaches correspond to theoretical perspectives and definitions of research problems within the programme. They are also a basis for systematizing and to some extent synthesizing the research. In more detail, the approaches are:

- *Natural resource and welfare economics.* An increased appreciation of the environmental resource base as a fundamental capital asset has led to important theoretical developments in economics, including a deeper understanding of the links between the economy and our natural resources and environments. We use an organizing principle based on social accounting matrices (SAMs). This improves scientific dialogue, promotes stakeholder involvement and becomes a natural way of disseminating and communicating results and ideas. A SAM can be constructed in many ways and each made to fit the purpose at hand. For example, the national accounts are organized as one particular SAM.
- *Natural resource ecology.* Adaptive resource management (ARM) offers a framework for addressing complexities and managing in the face of uncertainty (Willebrand et al, 2006). As with most biological systems, the ecological dynamics of the complex systems present in the mountain region are only partly understood. In most cases, observations of these processes also entail uncertainty. It may even be difficult to substantially reduce this uncertainty before the system has changed. Such uncertainty leads to a reduction in management precision, even when the objectives of management are agreed upon. Nevertheless, decisions must regularly be made despite the uncertainty. Ignoring outliers and putting a strong belief in stability, causality and predictive change may make the unexpected even more unexpected.
- *Democracy and public administration of natural resources.* A central issue is the tension between international and national policy-making and the demands for increased local management of natural resources. The key issue is how to design sustainable management regimes that fit local requirements and, at the same time, correspond with various international and national policies among diverging societal sectors. What are the consequences of different property rights allocations as well as different types of co-management practices? The central question is: who has the right to make decisions and with what legitimacy?

Phase two still has some months to go at the time of writing, and there are many exciting results on the way. There has been an emphasis on synthesis and reports directed to stakeholders at different levels. A large part of the work in phase two has been a nationwide survey that over-sampled the Mountain Mistra communes to make it possible to analyse the results at the level of a single commune. More than 11,000 questionnaires were sent out in a mail survey, each containing 20 pages, with more than 70 questions. The average response rate was well above 60 per cent, and the responses have already provided new knowledge regarding attitudes and values with regard to large carnivores, fish and wildlife, the right of common access and protected areas. To some extent, this has given support to the tension presented in Figure 8.2, for example, while a large proportion of the households in the mountain communes use meat from game on a regular basis, this is much more uncommon in the more populated south. The eight northernmost Mountain Mistra communes also show a much higher proportion of respondents who think that the protected area in the mountain region is too large.

There has been a maturing of several of the different modelling approaches in ecology and economy to provide tools for managers, and it is now possible to compare their strengths and weaknesses. These tools will be evaluated by stakeholder representatives using scenario techniques.

Conclusions

Stakeholder involvement

The Mountain Mistra programme is hosted by the Swedish Agricultural University, and many of the researchers involved had experience of working together with stakeholders in applied projects on forestry, fishery, wildlife management and so on. Partly using this experience, a full-time communicator with experience from the media was brought into the programme and was able to build a successful strategy (Esselin and Ljung, 2006). The fact that most of the programme board are stakeholders served to provide additional support and experience. We look upon this process as a two-way social interaction and learning process, enabling both researchers and users to understand key ecological, economic and social factors and their interdependencies. It is important that the stakeholders feel engaged and enthusiastic about the programme. This requires active participation and hard work from both the researchers and the stakeholders themselves. The stakeholders' understanding, insight and knowledge must be clearly communicated to the researchers in order to help them to adjust and develop the research in a way that takes the stakeholders' needs into consideration. At the same time, it is the stakeholders who can apply the scientific results in practice. The stakeholders in the user network come from state agencies, regional administrative boards, mountain communes, reindeer herders and NGOs representing forestry, hunting, fishing tourism and so on. The people in the network have gradually changed over time and the individuals do not always have an official mandate from the organizations they represent. It is important that they are free to express their own opinions and interact dynamically in the discussions.

The important key to success in involving stakeholders in the research process has been to build two-way trust through regular meetings. The process has evolved from researchers presenting ideas and results to a rather passive audience to workshops where small mixed groups of researchers and different stakeholders actively discuss selected topics. A communicator with a background from outside the academic world greatly helped to facilitate this process. Our goal is to build a network that will continue long after the research programme has finished. The annual Mountain Research Conference is now in the hands of the user network with support from the programme, and not the other way around.

The communicator regularly makes visits to many of the mountain communes to meet with different stakeholders. Ten times a year, we produce *FjällFokus* (Focus on the Mountains), a printed newsletter of 4–8 pages that summarizes research findings in a popular form. The members of the network receive *FjällFokus* and every month a personal email newsletter is produced. The programme also produces a formal annual report. Though this is primarily directed to Mistra, it has also become popular among the stakeholders, and its format is now based on suggestions from the users (for example, to be more thematic). The stakeholders also contribute to this report. The Mountain Mistra programme also has its own report series, which is mostly published in Swedish. *FjällFokus* and the report series can be downloaded from the website (www.mistra.org/program/fjallmistra/hem), which also provides current news, a calendar of meetings and up-to-date information about the programme, as well as relevant links to other websites for information about the management of the mountain region. Media cooperation helps to circulate results and findings from the programme to a broader audience. Press releases are sent out on a regular basis, and *FjällFokus* is distributed to the media covering the northern part of Sweden, as well as to national media.

Interdisciplinarity in the project team

In phase one, the Mountain Mistra programme was characterized by collaboration from a great number of research disciplines with common research goals. A significant amount of scientific effort was put into each of the seven research groups. Phase two could not have been structured successfully without the knowledge and experience from phase one. The research design in phase two provides for increased interaction across disciplines compared to the previous phase.

The number of scientists representing different disciplines is more-or-less equal, with a slight predominance of the social sciences. Also, within the social science team there is now a wider representation of disciplines compared with phase one, including economists, sociologists and a human geographer, as well as political scientists and a historian.

The executive group is comprised of three professors together with the programme director, and is responsible for the scientific management of the synthesis and interdisciplinary work. The executive group has a good interdisciplinary balance between the social and natural sciences, with two natural scientists (with broad expertise from ecology) and two social scientists (from natural resource economics and political science). Additionally, this executive group has the responsibility to develop the three theoretical frameworks described earlier. The four focal points facilitate interdisciplinary research because they require different research expertise, and we explicitly explored the enthusiasm and commitment of the scientists from phase one before they were invited to join phase two. Our emphasis on integration, synthesis and dissemination within and across every element of the programme places a large burden on each researcher, in addition to the traditional and equally important requirements for quality and quantity in scientific publishing.

Lessons for integrated research and management

The idea that traditional disciplinary science cannot solve the environmental problems of a sustainable future is becoming increasingly accepted. The obvious question to ask is: who should have the responsibility for the process of synthesizing piecewise research results into a comprehensive understanding of the process that needs management? Researchers in the scientific community are sometimes compared with the blind man feeling the elephant, or accused of making type III errors in their analysis – giving a correct answer but to a wrong question.

The problems of managing natural resources are of such complexity that they require integrated synthesis across several research disciplines, and close cooperation with the stakeholders who will implement the interdisciplinary synthesis (Willebrand et al, 2006). There is now a wide source of experience on how to implement this latter requirement. It is a matter of allocating resources to this process, but not only by redirecting research money. It is a two-way process, and stakeholders also have to make an effort to obtain a useful tool. In this two-way process, neither the research community nor the stakeholders have all the answers, and this certainly puts new demands on both groups. It is not only researchers who have discovered that environmental problems require more complex solutions than expected!

Interdisciplinary research is more difficult to manage than stakeholder integration. Most researchers will agree that finding a solution to many of the complex environmental problems faced by society requires an interdisciplinary approach. However, most disciplines rest on a general theory that is continuously advanced by disciplinary research; applied and interdisciplinary research must have a foundation in this general theory. The major driver of interdisciplinary research is the complex applied problems, not the need to develop new scientific theory by integrating disciplines. I would therefore caution against interdisciplinary PhD programmes since there is an obvious risk that they will become 'jack of all trades, master of none'. The university system is highly career-driven and, to a large extent, based on publication rate, which by definition reduces the interest in interdisciplinary research. Researchers in any joint integrated interdisciplinary research will have to adjust their effort towards a common goal and spend time and resources to understand the bigger picture. One solution is to specifically allocate resources to interdisciplinary research, and Mistra is a very successful example. The risk is that this will create researchers who have poor competitiveness in obtaining funding from the traditional research councils, and that joining an interdisciplinary programme may well become a dead end. However, I am certain that the Mountain Mistra researchers have acquired skills that are vital for the development of disciplinary science.

References

Aanes, S., Engen, S., Sæther, B.-E., Willebrand, T. and Marcstrom, V. (2001) 'Sustainable harvesting strategies of willow ptarmigan in a fluctuating environment', *Ecological Applications*, vol 12, pp281–290

Andersson, J. (2003) 'Effects of diet-induced resource polymorphism on performance in arctic char (*Salvelinus alpinus*)', *Evolutionary Ecology*, vol 5, pp231–228

Baskin, L. M. and Hjältén, J. (2001) 'Fright and flight behavior of reindeer (*Rangifer tarandus* L.)', *Alces*, vol 37, pp435–445

Bostedt, G., Parks, P. and Boman, M. (2002) *Integrating Forestry and Reindeer Husbandry in Northern Sweden: A Discrete Time Application*, Working paper 308, Umeå, Department of Forest Economics, Swedish University of Agricultural Sciences.

Bruun H. E., Moen J. and Angerbjörn A. (2003) 'Environmental correlates of meso-scale plant species richness in the province of Härjedalen, Sweden', *Biodiversity and Conservation*, vol 12, pp2025–2041

Byström, P., Andersson, J., Persson, L. and De Roos, A. M. (2003) 'Size-dependent resource limitation and foraging-predation trade offs: Growth and habitat use in young arctic char', *Oikos*, vol 104, pp109–121

Carlsson, J. (2000) *Microsatellite Variability in Swedish Arctic Charr* (Salvelinus alpinus *L.*), Masters thesis, Umeå, Swedish University of Agricultural Sciences

Ericsson, G., Sandström, C. and Bostedt, G. (2006) 'The problem of spatial scale when studying human dimensions of a natural resource conflict: Humans and wolves in Sweden', *International Journal of Biodiversity Science and Management*, vol 2, pp343–349

Eriksson, T., Andersson, J., Byström, P., Hörnell-Willebrand, M., Laitila, T., Sandström, C. and Willebrand, T. (2006) 'Fish and wildlife in the Swedish mountain regions: Resources, use and management', *International Journal of Biodiversity Science and Management*, vol 2, pp334–342

Esselin, A. and Ljung, M. (2006) 'Bridging the gap: The Mountain Mistra programme as an arena for collaborative learning', *International Journal of Biodiversity Science and Management*, vol 2, pp315–325

Fredman, P. and Emmelin, L. (2001) 'Wilderness purism, willingness to pay and management preferences: A study of Swedish mountain tourists', *Tourism Economics*, vol 7, pp5–20

Fredman, P., Emmelin, L., Heberlein, T. and Vuorio, T. (2001) 'Tourism in the Swedish mountain region', in Sahlberg, B. (ed.) *Going North*, Östersund, European Tourism Research Institute, p6

Fredman, P. and Heberlein, T. (2001) *Changing Recreation Patterns among Visitors to the Swedish Mountain Region 1980–2000*, Working paper, Östersund, European Tourism Research Institute

Gong, P., Boman, M. and Mattsson, L. (2001) 'Multiple-use forest planning techniques: A synthesising analysis', *Studia Forestalia Suecica*, vol 212, p27

Gong, P., Mattsson, L. and Boman, M. (2002) *Non-timber Benefits, Price Uncertainty, and Optimal Harvest of an Even-aged Stand*, Working paper, Umeå, Department of Forest Economics, Swedish University of Agricultural Sciences

Lindberg, K., Andersson, T. and Dellaert, B. (2001) 'Tourism development: Assessing social gains and losses', *Annals of Tourism Research*, vol 28, pp1010–1030

Lindberg, K., Denstadli, J., Fredman, P., Heldt, T. and Vuorio, T. (2001) *Skiers and Snowmobilers in Södra Jämtlandsfjällen: Are there Recreation Conflicts?*, Working paper, Östersund, European Tourism Research Institute

Moen, J. and Danell, Ö. (2003) 'Reindeer in the Swedish mountains: An assessment of grazing impacts', *Ambio*, vol 32, pp397–402

Moxnes, E., Danell, Ö., Gaare, E. and Kumpula, J. (2001) 'Optimal strategies for use of reindeer rangelands', *Ecological Modelling*, vol 145, pp225–241

Nordh, M. (2000) '*Modellstudie av potential för renbete anpassat till kommande slutavverkningar* [Simulation study of potential for reindeer grazing adapted to coming clear-cuts]', Arbetsrapport 77, Working paper, Umeå, Swedish University of Agricultural Sciences

Sandström, C., Moen, J., Widmark, C. and Danell, Ö. (2006) 'Progressing toward co-management through collaborative learning: Forestry and reindeer husbandry in dialogue', *International Journal of Biodiversity Science and Management*, vol 2, pp326–333

Skarin, A. (2001) *Interactions Between Reindeer, Humans, Topography and Weather: Spatial Patterns of Reindeer Pellet Groups and Lichen Height*, Graduate study thesis 221, Working paper, Uppsala, Department of Animal Breeding and Genetics, Swedish University of Agricultural Sciences

Tannerfeldt, M., Elmhagen, B. and Angerbjörn, A. (2002) 'Exclusion by interference competition? The relationship between red and arctic foxes', *Oecologia*, vol 132, pp213–220

Willebrand, T. and Hörnell, M. (2000) 'Understanding the effects of harvesting willow ptarmigan *Lagopus lagopus* in Sweden', *Wildlife Biology*, vol 7, pp205–212

Willebrand, T., Sandström, C. and Lundgren, T. (2006) 'Reaching for new perspectives on socio-ecological systems: Exploring the possibilities for adaptive co-management in the Swedish mountain region', *International Journal of Biodiversity Science and Management*, vol 2, pp359–369

Zachrisson, A., Sandell, K., Fredman, P. and Eckerberg, K. (2006) 'Tourism and protected areas: Motives, actors and processes', *International Journal of Biodiversity Science and Management*, vol 2, pp350–358

Zhou, W. and Gong, P. (2002) *Multiple-use Trade-offs in Swedish Mountain Region Forests*, Working paper 320, Umeå, Department of Forest Economics, Swedish University of Agricultural Sciences

9

Reconciling the Conservation of Biodiversity with Declining Agricultural Use in the Mountains of Europe: The Challenge of Interdisciplinary Research

Jonathan Mitchley, Joseph Tzanopoulos and Tamsin Cooper

Background

For centuries, agriculture has played an important role in sustaining Europe's rural environment and its biodiversity through the management of landscapes and habitats (Beaufoy et al, 1994; McCracken et al, 1995). The often harsh environmental conditions and the long history of farming in the uplands have resulted in an anthropogenically maintained dynamic equilibrium between various vegetation types, creating landscapes that are both 'open' and structurally diverse (Huber et al, 2005). As a result of significant recent agricultural adjustment, however, processes of agricultural contraction, decline or even abandonment are now widespread in the mountains of Europe (Baldock et al, 1997; MacDonald et al, 2000), with potentially major impacts on landscapes and mountain biodiversity. Deleterious impacts include landscape closure and increasing forest dominance (Debussche et al, 1996; Gordon et al, 1981; Naveh and Lieberman, 1984; Nelson, 1990), and the loss of locally adapted open-habitat species (Suarez-Seoane et al, 2002).

The abandonment of agricultural land, however, can initiate new successional pathways in ecosystems overexploited in the past, providing opportunities for 'rewilding' natural vegetation (Giourga et al, 1998) and for the reintroduction of some of the large predators that occurred in the pre-agricultural landscape, including raptors, wolves and bears (Breitenmoser, 1998). With shifts in the profile of rural inhabitants, land may be subject to alternative uses, such as hunting and rural tourism, supporting the preservation of traditional landscapes (Pinto-Correia and Mascarenhas, 1999).

For decades, policy-makers have attempted to address the problems in mountain areas through the EU's Less Favoured Area (LFA) Directive

(75/268), designed to secure the continuation of farming through income support in areas hampered by permanent natural handicaps, and thus to prevent land abandonment (Caraveli, 2000; Terluin et al, 1995). Different agricultural policy approaches make different assumptions about where the limits to acceptable change lie. Further, policies increasingly seek to conserve the countryside by linking biodiversity objectives directly to the viability of farming and to this end, biodiversity objectives have been integrated into the EU Common Agricultural Policy (CAP) through the Agri-Environment Regulation (2078/92) and Rural Development Regulation (1257/99).

Europe's mountain biodiversity, and the human communities that live and work among it, face unprecedented threats from social, economic and environmental forces of change (Huber et al, 2005). These same forces also bring exciting opportunities for the integration of knowledge and expertise to achieve sustainable solutions for the future development trajectories of these mountain areas. These are the issues central to the BioScene research project. The aim of this chapter is to outline the interdisciplinary aspects of the research methodology in BioScene and to draw down some lessons and recommendations for interdisciplinary research practice more generally.

Research aims, objectives and methodology

BioScene was funded by the European Commission's 5th Framework Programme for Research and Technological Development (RTD) under the Energy, Environment and Sustainable Development Programme. The project's overall aim was to evaluate the biodiversity consequences of agricultural decline and restructuring in mountain areas of Europe, and to provide recommendations and strategies for reconciling biodiversity conservation with the impacts of declining agricultural use within the context of agri-environmental, nature conservation and rural development policies, and the Sustainable Development Strategy and Impact Assessment Communication and Guidelines.

Carried out between 2002 and 2006, the research adopted a case-study approach to analyse the biodiversity consequences of different scenarios of agricultural change in six contrasting European countries (UK, France, Greece, Slovak Republic, Norway and Switzerland) (see Figure 9.1) covering the major biogeographical regions of Europe. For each study area, the objectives were:

- to identify the policy context and socio-economic drivers of change over the past 25 years (since the early 1980s);
- to describe the associated ecological changes (trends and issues for species and habitats);
- to conceptualize four alternative scenarios, or futures, extending over a 25-year period spanning 2005 to 2030, to explore the implications for

landscape and biodiversity of varying configurations of agricultural policy and market reform;

- to produce 'visualizations' to depict the expected land-use, landscape and biodiversity changes given the assumptions underpinning each scenario;
- to reveal local understandings of the nature of landscape change, the relationship between agriculture and biodiversity conservation through deliberative participation by regional stakeholders, and where the limits to acceptable future change lie in the context of agricultural decline;
- to assess the social acceptability, cost effectiveness and overall sustainability of the scenarios using a sustainability assessment (SA) methodology including discussions with stakeholders;
- to provide recommendations for integrating biodiversity conservation strategies with rural development policy and sustainable development.

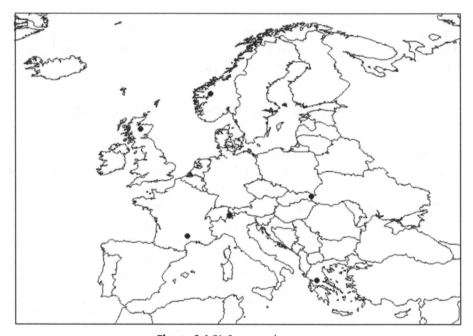

Figure 9.1 BioScene study areas

Note: The areas are Norway (Sjodalen valley, East Jotunheimen), Scotland (northeastern Cairngorms National Park), Switzerland (Mid Grisons), Slovakia (Bukovské vrchy mountains), France (Causse Méjan plateau) and Greece (Pindos mountains)

Scenarios have been deployed as heuristics: 'imaginative pictures of potential futures' (Penker and Wytrzens, 2005), that explore 'what might happen if...' (Veenelkaas and van der Berg, 1995) certain assumptions are made about the nature and direction of future change. Scenarios capture key ingredients of the uncertainty about the future of the system being studied (Peterson et al, 2003).

Scenario analysis is a flexible tool and can be used to support policy optimization, vision building or strategic orientation (Westhoek et al, 2006).

Within BioScene, we used scenario analysis to build strategic orientation in mountain areas by asking: what alternative futures are possible under alternative trajectories in the evolution of mountain landscapes? Of the four scenarios investigated in the project, the first, the Business as Usual Scenario, is a descriptive scenario that assumes an extrapolation of current trends; it is the baseline against which the implications for biodiversity and landscape of the other three scenarios can be compared and contrasted. It assumes the continuation of EU and/or state support for agriculture but that this will be insufficient to reverse the long-term trend of agricultural decline in mountain areas, resulting in some restructuring and a reorientation towards environmental stewardship and away from mono-productivism.

The other scenarios are exploratory and encapsulate three central ideas: neoliberalism, biodiversity management and natural ecological processes. These three scenarios are founded on the same premise – that agricultural use in mountain areas will decline – in order to explore the consequences of this process, to disrupt traditional constructions of the relationship between agriculture and biodiversity conservation, and to explore the ways in which different assemblages and profiles of biodiversity in the uplands could be derived.

The Agricultural Liberalization Scenario describes a future in which there is increasing political resolve to liberalize Europe's agricultural sector, such that all commodity support for agriculture is suddenly withdrawn, including funding formerly distributed under the EU Agri-Environmental (2078/92) and Rural Development (1257/99) Regulations. Under this scenario, the speed and extent of agricultural decline are accelerated, with a retraction in the area of land under agricultural management and the transfer of former agricultural land into other land uses and/or subject to different land management regimes.

The Managed Change for Biodiversity Scenario is also predicated on the premise that agricultural liberalization will occur, but under this scenario it is assumed that the state and other stakeholders decide to intervene in areas significant for biodiversity, to manage this process of restructuring to extract nature conservation dividends. As such, programmes are set up to halt the loss of priority species and habitats, resulting in the continuation of extensive agricultural practices, with the role of farmers reconfigured to one of biodiversity and landscape stewards.

The Wilding Scenario also assumes that agricultural liberalization proceeds and that this will lead to the exodus of some farmers and the liberation of former agricultural land. Under this scenario, however, the state and society decide not to continue to underwrite farming structures but instead to instigate a radical conservation programme aimed at re-establishing natural ecological processes at a landscape scale and accompanied by ecological restoration initiatives.

Interdisciplinarity in environmental research practice

As modern societies face the challenges posed by sustainable development and environmental issues, there is an increasing demand for integrated and interdisciplinary approaches to problem-solving and research including greater interaction between researchers, managers and decision-makers and producing policy relevant outcomes (Bruce et al, 2004; Lowe and Phillipson, 2006; Mattison and Norris, 2005; Watson, 2005; Wijkamn, 1999). Conventional academic disciplines organize knowledge around a particular world-view and have structured methods of research practice and inquiry. The world of scholarship has been broken up into disciplines to make scientific discovery and progress possible (Daily and Ehrlich, 1999). Such disciplinary specialization, and the resultant fragmentation of knowledge, however, limit the capacity to address issues of a more holistic and complex nature. Dealing with complexity requires approaches that transgress disciplinary boundaries and that build on cross-discipline communication, interaction and integration.

The research community, in response to the current challenges, has developed various approaches to cross-discipline research that fall into three main categories, although the precise definition of each often varies between commentators (see Balsiger, 2004; Bruce et al, 2004; Harvey, 2006; Tress et al, 2005b). Below we follow broadly the exposition by Bruce et al (2004) since this was developed in the context of an analysis of interdisciplinarity within the European 4th Framework Programme for RTD:

1   Transdisciplinary research, which involves the organization of knowledge around complex heterogeneous domains (cf. disciplines and subjects), creating new approaches tailored to the problem context rather than the academic discipline.
2   Multidisciplinary research, which is not based on problem-solving, but is thematically oriented, so that an issue is approached by a range of disciplines, with researchers from each working in a self-contained manner and from a clearly disciplinary perspective; although there may be linkages between disciplines, there is little synergy in the outcomes.
3   Interdisciplinary research, which literally means 'between disciplines' and, as such, approaches an issue from a range of disciplines, integrating them in order to exploit synergies and to generate an holistic outcome (Karlqvist, 1999).

Bruce at al (2004) further divide interdisciplinary research into:

•   Mode 1 research, which brings disciplines together in order to overcome an obstruction to further development within a discipline or to enable a discipline to move into new areas of research.

- Mode 2 research, where the primary aim is problem-oriented, to cut across disciplinary boundaries in the 'new production of knowledge' and, as such, renders discipline-related outputs less central.

Interdisciplinarity in BioScene

BioScene addresses a complex and context-specific set of science–society–environment issues: seeking to reconcile agricultural decline with biodiversity conservation within the frame of the sustainable development of Europe's mountainous areas. The research investigates the implications for landscape and biodiversity of agricultural decline through an examination of: first, declining agricultural activity in mountains; second, public perceptions of landscape change; third, the socio-economic drivers of change, their structural effects and land-use consequences; and fourth, the social acceptability, economic feasibility and impacts on biodiversity conservation of four scenarios.

Interdisciplinarity requires a comprehensive approach to problem-solving, viewing problems with a variety of methods and approaches and then working to assemble their partial insights into something approximating a composite whole. Within BioScene, a range of disciplines and modes of inquiry have been harnessed, including landscape ecology, restoration ecology, zoology, geography, agro-economics, sociology and environmental psychology. The research outcomes include recommendations for integrating conservation and sustainable rural development at a pan-European scale. Given the holistic and integrated nature of these strategies, it is clear that BioScene demanded an interdisciplinary 'Mode 2' approach.

Interdisciplinary research has the exciting potential to provide effective approaches to understanding complex issues. In comparison to the pursuit of monodisciplinary research, however, it entails greater risks and fewer rewards due to the barriers afforded by conventional institutional arrangements and academic assessment structures (Brewer, 1999; Bruce et al, 2004; Campbell, 2005; Daily and Ehrlich, 1999; Harvey, 2006; Naiman, 1999; Pickett et al, 1999; Tress et al, 2005a; Wear, 1999). Working across disciplines can often be problematic because:

- there may be difficulties in communication and understanding of the 'languages' employed within disciplines;
- there may be conflicts between theoretical and applied approaches;
- success rests on the researcher's ability to step outside their disciplinary world-view and to engage with different epistemological bases, to understand the different ways in which knowledge claims are verified and to recognize the existence of multiple criteria for assessing truth, reality and validity.

We anticipated these challenges within BioScene and, during the design of the research, special emphasis was given to the following:

- building a consortium of researchers with a positive attitude towards interdisciplinary research and, where possible, with experience in working across disciplines;
- scheduling regular meetings to encourage frequent interaction between social and natural scientists, including sessions devoted to general discussion of theoretical and methodological approaches in order to facilitate the exchange of knowledge, the cross-fertilization of ideas, and mutual understanding and acceptance;
- including, at the same time, distinct disciplinary work packages in the work programme to facilitate the production of disciplinary publications and other outcomes traditionally rewarded by the academic discipline structure, thus providing extra motivation for researchers to contribute fully to the project;
- encouraging integration of the ecological and socio-economic strands of the research project through a range of methodological tools, including scenarios, 'visualizations' and sustainability assessment;
- fostering a culture of mutual academic tolerance and flexibility, allowing adaptive management of the project's work structure while adhering to the overall aims and objectives of the research;
- adopting a consistent methodology across case studies combined with a degree of flexibility to allow adaptation in the light of country-specific contextual differences.

Stakeholders and sustainability assessment in BioScene

Stakeholders can play an important role as co-participants in an interdisciplinary research process because of the focus of such projects on environmental and sustainability issues, their relevance for public policy, and the associated challenge of effecting social and behavioural change (Berkhout et al, 2003). Within the frame of interdisciplinary research, stakeholders can be divided into two main categories: individuals, such as policy-makers and government officials who can be identified as having a 'stake' in the research project and its outcomes; and those who have a 'stake' in the topic under investigation, and consequently include a much broader cross-section of society.

The involvement of stakeholders at different stages in the project's life cycle may contribute significantly to more effective policy outcomes (Bruce et al, 2004). Stakeholders can have an important input in the development of the research design by directing the focus to issues relevant to the real world. Their input to the actual research, and their contributions in the form of discussions at stakeholder meetings, workshops and citizen panels, serves to challenge academic framings of the issue under study through the mobilization and application of lay knowledges (Burgess et al, 2000). Finally, stakeholder involvement in the assessment of the project's outputs is particularly important as the successful implementation of the recommendations often depends on the acceptance by stakeholders of these recommendations and their willingness, capacities and resources to implement them.

Stakeholder engagement in BioScene was facilitated through the establishment of stakeholder groups in each of the six study areas. These groups involved about 12 people local to the area who were identified following preliminary meetings with about 60 individuals. The aim was to constitute groups of people who held a broad range of views, to encourage a deliberation between people with different perspectives and to be as inclusive of as many 'voices' as possible (Bloomfield et al, 2001). As such, individual stakeholders were seen as 'speakers for' competing discourses that circulate in the study area about the nature of landscape change (see O'Neill, 2001), with different understandings of past changes and with alternative visions for the future. We were also mindful of achieving an appropriate gender mix and of including people with a range of ages, given the research focus on past and future processes of change.

Three stakeholder meetings were scheduled during the course of the three-year BioScene project in each study area. The first meeting provided partners with an insight into stakeholders' perceptions, assumptions and expectations for the future of their land and for their livelihoods, allowing them to identify narratives of landscape change. This material, together with interviews and an analysis of policies for agricultural change (local, national and European), was instrumental to the first stages of the SA process: the definition of study area sustainability objectives and of the sustainability baseline reference document identifying major issues, trends and challenges for each area. This first meeting was conducted at the end of the first year of the project and group members were encouraged to discuss, in their own terms, past processes of change and to envision future landscape changes, to allow the researchers to explore the ways in which different trajectories are advocated and preferences defended.

From this first meeting, a capital of understandings about landscape evolution, the nature of landscape change and the potential future changes, was fostered within each group. Thus, at the second meeting, held at the end of year two, stakeholders were presented with the four scenarios described above, and manipulated photographic representations – 'visualizations' – of the possible landscape and biodiversity outcomes, for a discussion on where the limits to acceptable future change may lie. The second meeting was designed to obtain the views of stakeholders through a visualization exercise. This was based on the biodiversity consequences of the three 'BioScenes' 'predicted' through the ecological modelling – of shifts in priority species and habitats, and landscape changes – which were translated (and synthesized) into photographic representations to stimulate feedback from the stakeholder groups concerning the acceptability of the 'BioScenes' presented to them. The discussions revealed stakeholder understanding of the scenarios and the causal relationships they represent, and explored where the limits of acceptable change might lie. They helped reveal how people understand and perceive the concept of biodiversity, knowledge about biodiversity conservation, agricultural practices and the interrelationships in mountain areas.

This second meeting also invited stakeholders to discuss the preliminary list of sustainability objectives for their areas compiled by partners with the aim of identifying any objective that raised serious disagreement, and to provide partners with additional objectives or themes that they might have missed but which were important to some of the stakeholders. Finally stakeholders were asked to identify three objectives (from lists of about 20), which they felt were the most important. Subsequently, the local partners and – independently – the authors, completed the SA matrix-based analysis of the scenarios, comparing these against the set of sustainability objectives. At this stage the matrices reflected the views of 'experts'.

Finally, the third meeting, held six months before the end of the project, was devoted to discussing the acceptability of the scenarios, the choices and trade-offs they imply, and the degree to which they can be considered a 'sustainable future'. At these final meetings and as a part of the SA process, partners presented the stakeholders with summary matrices showing the results of the SA for all three scenarios. The perspective of 'civil society' was incorporated into the SA process through the presentation and discussion of the preliminary results of the matrix analysis with stakeholders during this final meeting.

Conclusion: Towards best practice in interdisciplinary research

Further details of our research methodologies and results are reported elsewhere (for example, Mitchley et al, 2006); the purpose here is to reflect on some of the lessons of BioScene for interdisciplinary research practice. Collaboration between natural and social scientists is increasingly being seen as vital for addressing complex socio-political issues inherent in developing strategies for sustainable development (Daily and Ehrlich, 1999; Lowe and Phillipson, 2006; Marzano et al, 2006). The most critical environmental issues underpinning sustainability involve concepts and issues that cross traditional scientific disciplinary boundaries and involve actors and stakeholders multisectorally. BioScene has confirmed the necessity for natural and social sciences to link their approaches and work together on environmental problems.

BioScene was conceived as an interdisciplinary project that aimed to build on the contributions of ecologists, economists, sociologists and human geographers. For this reason, each country team included members from a range of these disciplines who were responsible for three strands (work packages) – socio-economic and policy analysis, ecological analysis and modelling, and SA

– and the analytical tasks, such as the different assessments, required close collaboration and joint outputs from all members. Of the three strands, SA was the one with the greatest potential to promote the interdisciplinary ethos in the project.

During BioScene, problems and difficulties surfaced that revealed fundamental epistemological differences between researchers, as well as the challenges of such a wide range of cultural backgrounds. However, the clear impression during the final project meeting (Conference on Biodiversity Conservation and Sustainable Development in Mountain Areas of Europe: The Challenge of Interdisciplinary Research, 20–24 September 2005, in Ioannina, Greece) was that major progress was made and significant learning across disciplines had been realized. This achievement represents research benefits reaching well beyond BioScene to future scientific research and collaboration networks across Europe and even globally. This impact represents an important example of European added-value in interdisciplinary research practice.

Significantly, those individuals who had been most sceptical about interdisciplinarity and the SA process were those who were often most convinced of the benefits of a new approach to investigation. For example, one ecologist partner raised the following strengths and weaknesses based on the SA experience: she appreciated the need for logical thinking along causal chains and the fact that the SA helped her team to identify possibilities and limitations of the area's situation within a social framework, and that this *may* have led to a certain empowerment of citizens. SA was seen as a tool to realize and balance completely different factors with a sustainability focus, demanding an interdisciplinary mindset. On the negative side, she felt the exercise was extremely time-consuming and raised concerns about the obstacles to interdisciplinary work given that it does not lead to any academic rewards, especially for natural scientists.

Lessons for interdisciplinary research

BioScene used interdisciplinary research tools such as scenarios, integrated landscape modelling and SA. BioScene emphasizes the value of participatory approaches and stakeholder groups as a method for appraising scenarios based on changes in land-use policies. Scenarios can be used as a way of integrating social, economic and natural sciences, and as a tool for approaching conflict issues with stakeholders, enabling scientists and stakeholders to reflect upon the problems in a more open-minded and fruitful way. Scenarios are an integrating tool that can be used by natural and social scientists and stakeholders to consider the best responses in a changing world fraught with multiple unknowns (Sutherland, 2006).

The researchers in BioScene were unanimous in their conviction that interdisciplinary research is very important, including those previously unfamiliar with interdisciplinary approaches or those who were sceptical at the outset and during the research. However, BioScene emphasized that interdisciplinary research remains a challenge at a number of levels. A great deal of time must be dedicated at the beginning of the project to interdisciplinary communication: developing a common language and vocabulary, discussing epistemological issues, even a crash course in relevant disciplinary knowledge (for example, ecological modelling, narratives). It is important to be prepared to work hard to make interdisciplinary research work (Marzano et al, 2006).

Finally, since academic livelihoods are increasingly based on the results of scrutiny from the peer-review process, BioScene has shown that joint papers across disciplines/countries/institutes should be initiated early on, otherwise such papers are never going to be written (Tress et al, 2005b). This is important because more published papers detailing the challenges of, approaches to, and outputs from interdisciplinary research are needed if we are to enhance the awareness, quality and policy-relevant outputs that such research activity can achieve (Harvey, 2006).

Finally, the challenges of interdisciplinary research for the scientific community are much broader and deeper than the academic publishing potential; the furtherance of effective interdisciplinary research requires the development of a dialogue between researchers, funding agencies, policy and decision-makers on interdisciplinary objectives, problems and solutions (Tress et al, 2005c).

Acknowledgements

BioScene was funded by the European Commission's 5th Framework Programme for Research and Technological Development under the Energy, Environment and Sustainable Development Programme, project number EVK2 2001 00354. We thank Martin Sharman for support and encouragement. J. M. thanks all partners in BioScene for their dedication, enthusiasm, commitment and imagination over the years.

References

Baldock, D., Beaufoy, G., Brouwer, F. and Godeschalk, F. (eds) (1997) *Farming at the Margins: Abandonment or Redeployment of Agricultural Land in Europe*, The Hague, IEEP London/Agricultural Economics Research Institute

Balsiger, P. W. (2004) 'Supradisciplinary research practices: History, objectives and rationale', *Futures*, vol 26, pp407–421

Beaufoy, G., Baldock, D. and Clark, J. (1994) *The Nature of Farming: Low Intensity Farming Systems in Nine European Countries*, London, Institute for European Environmental Policy

Berkhout, F., Leach, M. and Scoones, I. (2003) 'Shifting perspectives in environmental social science', in Berkhout, F., Leach, M. and Scoones, I. (eds) *Negotiating Environmental Change: New Perspectives from Social Science*, Cheltenham, Edward Elgar

Bloomfield, D., Collins, K., Fry, C. and Munton, R. (2001) 'Deliberation and inclusion: Vehicles for increasing public trust in UK public governance?', *Environment and Planning C: Government and Policy*, vol 19, pp501–513

Breitenmoser, U. (1998) 'Large predators in the Alps: The fall and rise of man's competitors', *Biological Conservation*, vol 83, pp279–289

Brewer, G. D. (1999) 'The challenges of interdisciplinarity', *Policy Sciences*, vol 32, pp327–337

Bruce, A., Lyall, C., Tait, J. and Williams, R. (2004) 'Interdisciplinary integration in Europe: The case of the Fifth Framework programme', *Futures*, vol 36, pp457–470

Burgess, J., Clark, J. and Harrison, C. (2000) 'Knowledges in action: An actor network analysis of a wetland agri-environment scheme', *Ecological Economics*, vol 35, pp119–132

Campbell, L. M. (2005) 'Overcoming obstacles to interdisciplinary research', *Conservation Biology*, vol 19, pp574–577

Caraveli, H. (2000) 'A comparative analysis on intensification and extensification in Mediterranean agriculture: Dilemmas for LFA policy', *Journal of Rural Studies*, vol 16, pp231–242

Daily, G. C. and Ehrlich, P. R. (1999) 'Managing Earth's ecosystems: An interdisciplinary challenge', *Ecosystems*, vol 2, pp277–280

Debussche, M., Escarre, J., Lepart, J., Houssard, C. and Lavorel, S. (1996) 'Changes in Mediterranean plant succession: Old-fields revisited', *Journal of Vegetation Science*, vol 7, pp519–526

Giourga, H., Margaris, N. S. and Vokou, D. (1998) 'Effects of grazing pressure on succession process and productivity of old fields on Mediterranean islands', *Environmental Management*, vol 2, pp589–596

Gordon, M., Guillerm, J. L., Poissonet, J., Poissonet, P., Thiault, M. and Trabaud, L. (1981) 'Dynamics and management of vegetation', in Di Castri, F., Goodall, D. W. and Specht, R. L. (eds) *Mediterranean-Type Shrublands*, Ecosystems of the World Vol 11, Amsterdam, Elsevier, pp317–344

Harvey, D. R. (2006) 'RELU special issue: Editorial reflections', *Journal of Agricultural Economics*, vol 57, pp329–336

Huber, U. M., Bugmann, H. K. M. and Reasoner, M. A. (eds) (2005) *Global Change and Mountain Regions: An Overview of Current Knowledge*, Dordrecht, Springer

Karlqvist, A. (1999) 'Going beyond disciplines: The meanings of interdiscplinarity', *Policy Sciences*, vol 32, pp379–383

Lowe, P. and Phillipson, J. (2006) 'Reflexive interdisciplinary research: The making of a research programme on the rural economy and landuse', *Journal of Agricultural Economics*, vol 57, pp165–184

MacDonald, D., Crabtree, J. R., Wiesinger, G., Dax, T., Stamou, N., Fleury, P., Gutierrez Lazpita, J. and Gibon, A. (2000) 'Agricultural abandonment in mountain areas of Europe: Environmental consequences and policy response', *Journal of Environmental Management*, vol 59, pp47–69

Marzano, M., Carss, D. N. and Bell, S. (2006) 'Working to make interdisciplinarity work: Investing in communication and interpersonal relationships', *Journal of Agricultural Economics*, vol 57, pp185–197

Mattison, E. H. A. and Norris, K. (2005) 'Bridging the gaps between agricultural policy, land-use and biodiversity', *Trends in Ecology and Evolution*, vol 20, pp610–616

McCracken, D. I., Bignal, E. M. and Wenlock, S. E. (eds) (1995) *Farming on the Edge: The Nature of Traditional Farmland in Europe*, Peterbourough, Joint Nature Conservation Committee

Mitchley, J., Price, M. F. and Tzanopoulos, J. (2006) 'Integrated futures for Europe's mountain regions: Reconciling biodiversity conservation and human livelihoods', *Journal of Mountain Science*, vol 3, pp276–286

Naiman, R. J. (1999) 'A perspective on interdisciplinary science', *Ecosystems*, vol 2, pp292–295

Naveh, Z. and Lieberman, A. S. (1984) *Landscape Ecology: Theory and Application*, New York, Springer-Verlag

Nelson, M. (1990) 'Abandoned farmland in France', *Landscape and Urban Planning*, vol 18, pp229–233

O'Neill, J. (2001) 'Representing people, representing nature, representing the world', *Environment and Planning C: Government and Policy*, vol 19, pp483–500

Penker, M. and Wytrzens, H. K. (2005) 'Scenarios for the Austrian food chain in 2020 and its landscape impacts', *Landscape and Urban Planning*, vol 71, pp175–189

Peterson, G. D., Cumming, G. S. and Carpenter, S. R. (2003) 'Scenario planning: A tool for conservation in an uncertain world', *Conservation Biology*, vol 17, pp358–366

Pickett, S. T. A., Burch, W. R. and Morgan Grove, J. (1999) 'Interdisciplinary research: Maintaining the constructive impulse in a culture of criticism', *Ecosystems*, vol 2, pp302–307

Pinto-Correia, T. and Mascarenhas, J. (1999) 'Contribution to the extensification/intensification debate: New trends in the Portuguese montado', *Landscape and Urban Planning*, vol 46, pp125–131

Suarez-Seoane, S., Osborne, P. E. and Baudry, J. (2002) 'Responses of birds of different biogeographic origins and habitat requirements to agricultural land abandonment in northern Spain', *Biological Conservation*, vol 105, pp333–344

Sutherland, W. J. (2006) 'Predicting the ecological consequences of environmental change: A review of the methods', *Journal of Applied Ecology*, vol 43, pp599–616

Terluin, I. J., Godeschalk, F. E., von Meyer, H., Post, J. H. and Strijker, D. (1995) 'Agricultural income in less favoured areas of the EC: A regional approach', *Journal of Rural Studies*, vol 11, pp217–228

Tress, B., Tress, G. and Fry, G. (2005a) 'Researchers' experiences, positive and negative, in integrative landscape projects', *Environmental Management*, vol 36, pp792–807

Tress, G., Tress, B. and Fry, G. (2005b) 'Clarifying integrative research concepts in landscape ecology', *Landscape Ecology*, vol 20, pp479–493

Tress, B., Tress, G. and Fry, G. (2005c) 'Integrative studies on rural landscapes: Policy expectations and research practice', *Landscape and Urban Planning*, vol 70, pp177–191

Veenekaas, F. R. and van der Berg, L. M. (1995) 'Scenario building: Art, craft or just a fashionable whim?', in Schoute, J. F. T., Finke, P. A., Veenekaas, F. R. and Wolfert, H. P. (eds) *Scenario Studies for the Rural Environment*, London, Kluwer, pp11–13

Watson, R. T. T. I. (2005) 'Turning science into policy: Challenges and experiences from the science-policy interface', *Philosophical Transactions of the Royal Society B: Biological Sciences*, vol 360, pp471–477

Wear, D. N. (1999) 'Challenges to interdisciplinary discourse', *Ecosystems*, vol 2, pp299–301

Westhoek, H. J., van den Berg, M. and Bakkes, J. A. (2006) 'Scenario development to explore the future of Europe's rural areas', *Agriculture, Ecosystems and Environment*, vol 114, pp7–20

Wijkman, A. (1999) 'Sustainable development requires integrated approaches', *Policy Sciences*, vol 32, pp345–350

10

Cumulative Effects Assessment: An Approach for Integrated Research and Management in North America's Crown of the Continent Ecosystem

Michael Quinn, Guy Greenaway and Danah Duke

Introduction

Cumulative effects and wicked problems

The most significant environmental issues currently facing society (for example, climate change, loss of biodiversity and habitat fragmentation) are the result of cumulative effects – the complex additive and synergistic effects in time and space of repeated and multiple actions (Hegmann et al, 1999; Shoemaker, 1994). Cumulative effects arise from a variety of situations and activities, such as large projects that produce significant environmental change, to numerous small, individually insignificant projects that in combination have a compounding and degrading effect on the environment (Kennett, 1999; Piper, 2002). Cumulative effects are pernicious because each individual action, when considered in isolation, may seem insignificant. However, the interaction of these effects in time and space results in highly complex and often unpredictable changes to ecological and social systems. Frameworks for understanding cumulative environmental effects (cumulative effects assessment) generally consider the sources, pathways and effects of environmental change (Cocklin et al, 1992a, 1992b; Contant and Wiggins, 1991).

Large-scale cumulative effects are phenomena arising from constellations of linked and dynamic problems embedded in the fabric of the communities in which they occur (Kreuter et al, 2004). In their seminal paper 'Dilemmas in a general theory of planning', Rittel and Webber (1973) classified such problems as 'wicked'. Wicked problems differ from 'tame problems' in that the latter may be highly complicated, but the problem can be clearly articulated and solved using appropriate and standard methods. In contrast, wicked problems defy simple problem definition and there may be considerable disagreement among stakeholders (often with differing value orientations) about what the problem is. Wicked problems are not fully understood until a solution has been developed. Furthermore, attempts to design a solution to a wicked problem may actually change the problem. Conventional, linear and causal chain approaches to planning and management are the antithesis of design methodologies required to deal with wicked problems. Addressing wicked problems requires design and systems approaches that commence with the acknowledgement that there is no single, right solution and that resolution is as much a social and political process as it is a scientific endeavour.

The proliferation of cumulative effects is partly attributable to the incremental and disjointed nature of decision-making – or tyranny of small decisions – that characterizes many contemporary institutional structures and regulatory approval processes (Creasey, 2002; Kahn, 1966). The command-and-control paradigm of fragmented natural resource management masks the true nature of wicked problems and exacerbates cumulative effects. Conventional natural resource management is characterized by jurisdictional fragmentation, with each administrative authority making decisions in isolation from the others. Furthermore, within each of the administrative boundaries, individual resource sectors are regulated in a piecemeal fashion with a focus on mitigation of local short-term effects. Each new subdivision, gas well site, cutblock or recreational trail is subject to an approval process that provides few requirements for consideration of cumulative effects (Kennett, 1999). Sustainable development, adaptive ecosystem-based management and the emergence of other holistic planning, policy and management paradigms echo the need for novel approaches to strategically address these cumulative effects in order to achieve the goals of ecological integrity, economic sustainability and social equity (Cocklin et al, 1992a, 1992b; Noble, 2002; Piper, 2002; Prato, 2003; Slocombe, 1994; Stinchcombe and Gibson, 2001).

The theory and practice of considering cumulative effects in the assessment of individual project proposals have advanced significantly over the past two decades (Baxter et al, 2001; Damman, 2002; Griffiths et al, 1998; Ross, 1998). The experience and insight gained from project-based cumulative effects assessment (CEA) points to the need for the consideration of cumulative effects at a more strategic level (Brown and Therivel, 2000; Drouin and LeBlanc, 1994; Marsden, 2001). Higher-level assessment of cumulative effects could result in the availability of more and better strategic information to aid in planning and to provide a more comprehensive context for the assessment of individual project proposals. However, the assessment of cumulative effects of policies, plans and programmes over large geographic areas and long timeframes requires the development of new approaches to addressing wicked problems that are both integrated and interdisciplinary (Dube, 2003; Kennett, 1999). Demonstration research that employs 'design judgment' rather than 'solution seeking logic' (Nelson and Stolterman, 2003) is an effective way to partner researchers with planners and managers to advance both the theory and the practice of integrated resource planning and management. This chapter presents the preliminary results of such a project.

Project area

The Crown of the Continent Ecosystem (CCE) is one of North America's most ecologically diverse and jurisdictionally fragmented ecosystems (Long, 2002). Encompassing the shared Rocky Mountain region of Montana, British Columbia and Alberta, this 42,000km² ecological complex spans two nations, one state and two provinces, as well as numerous aboriginal lands, municipal authorities, public land blocks, private properties and working and protected landscapes (see Figure 10.1).

The headwaters of three of North America's major river systems, the Saskatchewan, Missouri and Columbia, flowing respectively to the Arctic, Atlantic and Pacific Oceans, are encompassed within the CCE. Internationally recognized for its ecological and geological uniqueness, the region is a continental nexus that co-mingles species and biotic communities from the prairie, mountain, boreal forest and coastal maritime regions. The ecological significance of the region is perhaps best indicated by the occurrence of eight large carnivore species and their associated prey; such an intact assemblage exists nowhere else in the lower 48 states of the US (see also Mattson and Merrill, this volume). The valleys of the CCE serve as important wildlife movement corridors, connecting meta-populations of various species throughout the Rocky Mountain cordillera. Many small mammals, birds, reptiles, amphibians, fish and a wide diversity of plants also contribute to the ecological importance of the CCE.

The CCE is currently facing an increase in human activity, primarily manifested by urban and rural residential expansion, recreation and resource extraction. Population growth rates in the region are among the highest in North America (Travis et al, 2002). Much of the growth is related to the high amenity values that are associated with the 'Rocky Mountain West' (Hansen et al, 2002). Ironically, the rapid growth and associated landscape change are eroding the very features that have made the region so attractive to newcomers. The growing realization that 'natural capital' must be maintained in order to preserve the value of these high amenity-value landscapes has led to calls for more comprehensive and interdisciplinary approaches to land-use policy, planning and management.

Regional landscape assessment project

In February 2001, government representatives from more than 20 agencies from across the transboundary CCE convened a meeting to explore ecosystem-based ways of collaborating on shared issues. Participants included representatives of

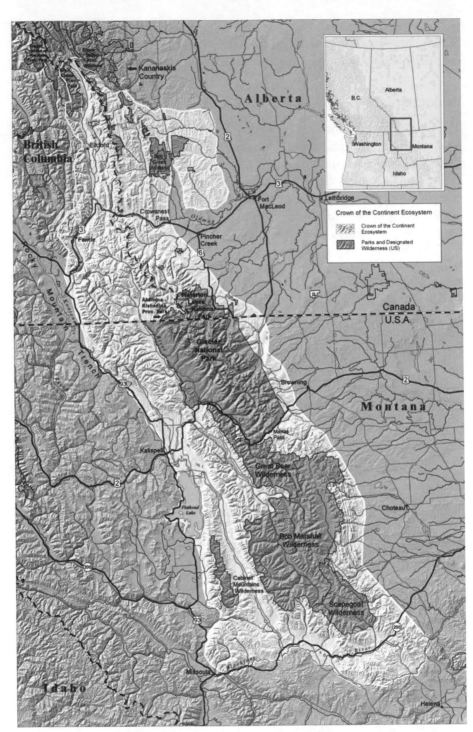

Figure 10.1 The Crown of the Continent Ecosystem

federal, aboriginal, provincial and state agencies or organizations with significant land or resource management responsibilities in the CCE. The aim was to bring together senior and middle managers with technical and professional staff who have a role in management at the ecosystem scale (for example, conservation biologists, land-use planners). The Miistakis Institute for the Rockies, a non-profit research organization affiliated with the University of Calgary, was invited to help facilitate the process and act as a neutral third party. Recognizing the potential of research partnerships, the organizers also invited academic partici-pants from the Universities of Montana and Calgary. The highly successful founding workshop, hosted by the Waterton-Glacier International Peace Park, resulted in a commitment by all participants to move forward collaboratively on regional management through the establishment of the Crown Managers' Partnership (CMP). The objectives of the CMP are to:

- build awareness of common interests and issues in the CCE;
- build relationships and opportunities for collaboration across mandates and borders;
- identify collaborative work already under way and opportunities for further cooperation.

The founding meeting also resulted in consensus around five strategic issues of importance to the participants. Emphasis was placed on the identifi-cation of issues that are best addressed at the larger regional scale, including:

1   addressing cumulative effects of human activity across the region;
2   addressing increased public interest in how lands are managed and how decisions are reached;
3   addressing increased recreational demands and increased visitation;
4   collaboration in sharing data and standardizing assessment and monitoring methodologies;
5   addressing the maintenance and sustainability of shared wildlife popula-tions. This chapter discusses progress on the first of these priorities.

Together with the CMP, the Miistakis Institute for the Rockies and the University of Calgary initiated the Regional Landscape Analysis Project (RLAP) to generate a collaborative approach to assessing cumulative effects in the transboundary CCE. Unlike other regional frameworks that have focused on the impacts the development of a single large resource base (Damman, 2002), the current framework integrates the effects of multiple land uses and natural disturbance regimes across a landscape using a multi-stakeholder and multi-sectoral, iterative approach.

In contrast to project-level CEAs that are largely reactive exercises, the framework addresses cumulative effects early in the decision-making process, allowing proposed developments to be assessed in relation to an overall landscape plan. This approach engages regional stakeholders in collaborative visioning, identification of barriers and drivers to advancing cumulative effects, acceptable threshold establishment, shared data management and modelling exercises.

The overall goal of the RLAP is to develop an educational and analytical tool that will assist land managers in developing a strategic framework to consider the cumulative effects of current and future land uses in the region. The objectives of the project are to:

- identify drivers and barriers to develop a multi-jurisdictional, international framework for regional CEA;
- develop a process to facilitate ongoing, collaborative data collection and harmonization for modelling regional cumulative effects;
- assess the value of both the process and model and make recommendations regarding the ongoing development of the framework and its implementation.

The size and complexity of the task require long-term commitment to the process. The findings reported herein should therefore be considered a work-in-progress.

Interdisciplinarity

The CMP includes participants from federal, state/provincial and aboriginal agencies in charge of agriculture, forestry, resource management, land, environment, parks, wildlife and other sectors, as well as academics. Participants come to the process with backgrounds in a wide variety of natural and social sciences. The design methodology being developed for the project is, by definition, interdisciplinary and design-based.

The primary mechanism for fostering interdisciplinarity is the comprehensive modelling process. The model adopted for the project (see below) enables the exploration and quantification of the effects of multiple human land uses and natural disturbance regimes on the entire region. The land uses that will be modelled and integrated are: agriculture, energy, transportation, human settlement, forestry, protected areas and tourism. The model helps to integrate human activities and biophysical responses and then link this information to regional economic models.

Land managers in the CCE have recognized the importance of making land-use decisions within a regional cumulative effects framework, however, no such framework is currently available in the region. Political, financial and technical barriers impede landscape-level collection of the information necessary for cumulative effects modelling. No single agency has the mandate or the resources to focus upon the entire region.

Interdisciplinarity requires much more than bringing disparate disciplines together (that is, multidisciplinarity); it requires the formulation of 'new ways of doing business', including holistic and cooperative thinking and action, with the participating disciplines informing each other's practice (Dickens, 2003; Turner and Carpenter, 1999; Wear, 1999). New models of practice require patience and understanding. This is difficult in a world of shrinking budgets, growing crises and impending deadlines.

Project outcomes

Drivers and barriers survey

A web-based survey was administered to members of the CMP in an effort to determine the level of knowledge and interest regarding CEA and to identify the most significant drivers and barriers to pursuing a complex regional CEA. A list of 65 potential survey participants was developed from the CMP membership list. Seven individuals from the initial list were not available for participation. From the remaining pool of 58 potential respondents, completed surveys were received from 36 individuals (effective response rate of 62 per cent). The 36 respondents identified themselves as being from Alberta provincial agencies (nine individuals or 30 per cent), Montana state agencies (three or 10 per cent), US federal agencies (six or 20 per cent), Canadian federal agencies (five or 16.7 per cent) and other institutions (for example, NGOs and universities) (seven or 23.3 per cent). There were no responses from British Columbia provincial agencies. Survey respondents were generally supportive of pursuing a regional CEA programme and recognized the value of such an initiative to their individual mandates.

Survey participants clearly indicated that the growth of the region's human populations and concomitant increases in human-use pressures were significant issues, both now and for the future. The top five most significant resource management issues identified were landscape connectivity, access management, recreation/tourism pressure, increasing linear disturbance and rural residential subdivision. Respondents articulated the need to address all of these issues across jurisdictional boundaries.

Participants indicated that the most important characteristics of effective collaborative CEA initiatives were:

- clearly defined and shared goals and objectives;
- shared commitment for long-term involvement;
- adequate commitment of resources;
- common issues and pressing need for response;
- frequent and effective communication;
- mutual respect and trust among participants.

Conversely, participants identified the most important barriers to effective collaboration initiatives in resource management as:

- lack of resources;
- lack of shared agency mandates and philosophies;
- lack of agency support; fourth, interagency barriers and 'turf' issues;
- lack of continuity of participating members.

Cumulative effects assessment framework

The fundamental components of the CEA framework being designed for this project include project management (includes shared issues identification), data collection, base case modelling, scenario modelling and the development of communication products for decision support. The process is iterative and adaptive, with the goal of continuous improvement.

The CEA framework being developed by the CMP includes the use of the ALCES® (A Landscape Cumulative Effects Simulator) computer model (Forem Technologies, 2006). ALCES® enables resource managers, industry, society and the scientific community to explore and quantify the cumulative, dynamic effects of human land-use practices and existing natural disturbance regimes. ALCES® contributes to a strategic cumulative framework through its use as an exploratory tool to identify emerging regional issues and opportunities, and by examining the potential implications of trends and policy choices under a range of future scenarios. The model is driven through a collaborative visioning process that ultimately contributes to planning regional sustainability. The benefits of building collaborative institutions, predicated on a shared understanding of issues and a high level of communication, extend well beyond the model results.

Modelling complex phenomena at the landscape level is data-intensive. There are three types of data required for ALCES® modelling: spatial, metric and trend. Spatial data exist in a variety of formats and scales and thus create challenges of harmonization across the study area. Metric data (non-spatial, non-trend parameters describing land use) and trend data (projections of future trends) describe and characterize the 'flows' – the factors that influence the changes in those stocks. As with spatial data, values of metric and trend data vary tremendously across the study area, are neither uniformly collected nor standardized, and (in the case of trend data) are conjectural. In many cases, the data simply do not exist in a published format. This requires convening workshops of subject experts who can work through a consensus-based process of determining, vetting and substantiating all metric and trend values.

Recommendations

Maintaining multiple agency involvement in a complex, multi-year project has been difficult, despite high support being repeatedly expressed by those agencies for a regional cumulative effects project. Agencies are operating under challenging conditions characterized by tight budgets, shifting priorities, changing governments and limited human resources. The result is that many of the participants have been hard-pressed to provide funding, time and personnel. However, a core group of participating agencies has continued to provide the necessary support for the initiative. At the time of writing (September 2006), the CMP had decided to take a slower approach to developing the RLAP, as discussed below. The use of other modelling approaches and alternatives from smaller-scale pilot projects are currently being assessed.

The following recommendations arise from the survey and from the authors' experience with the CMP. They are not in any particular order of priority, but are considered to be the most important identified by this project.

Higher-level support

The CMP should continue to explore the potential mechanisms and appropriate timing for attaining higher-level recognition and support for the partnership and the cumulative effects project. It is clear from the results of the current research that, without a stronger commitment for financial and human resource support, the cumulative effects project is not likely to be sustainable. The CMP currently operates without any formal mandate from the federal, provincial/state or other levels of government. It may be necessary to seek a formal mandate for the CMP that includes a multilateral agreement between the participants.

Clear articulation of project goals and objectives

Although the cumulative effects project has already commenced, the results of our research suggest that the members of the CMP lack a clear understanding and shared vision of project goals and objectives. The cumulative effects model and approach have been presented at the annual forums attended by the CMP membership, but this is evidently not enough to adequately create the level of engagement required for meaningful participation of all members. Participating agencies all have different needs and statutory requirements for CEA that impede the ability to conduct a regional process. Furthermore, as with most wicked problems, there is considerable divergence on agreeing what problems are being addressed by the project.

Shared approach to dedicated resource allocation

The potential to leverage contributions through a multi-agency initiative, such as the CMP, is a strong incentive for participation. For example, a US$5000 annual agency commitment to the CMP, multiplied by 20 participating agencies, results in US$100,000 of core support. The programme facilitator, in this case the Miistakis Institute, can then seek matching contributions from other government and private sector sources. However, the ability for agencies to commit resources to a transboundary, international initiative creates administrative challenges that must be overcome.

Development of internal communication products

Communication with CMP participating agencies must be increased between the annual forums. Furthermore, materials should be developed and distributed to individuals beyond the delegates to the annual meetings. Individual agency representatives should be contacted to develop a list of targets to receive specific communications. The CMP is in the process of finalizing a website and a suite of presentations that can be used by participating agencies to communicate the activities of the partnership (www.rockies.ca/cmp). Communication products should focus on issues of shared interest, the benefits of collaboration and the demonstrable outcomes of CMP initiatives.

Development of external communication products

To date, the CMP has operated relatively quietly with little external outreach to the general public or to potentially interested non-governmental interest groups. Now that the initiative is well established, it seems appropriate to communicate the ideals and benefits beyond the current participants. Such public outreach may also help to develop the higher-level support identified above. Generating an increased understanding of the CMP may also result in expanded participation in its activities.

Monitoring, feedback and continuous improvement

The CMP should ensure that there are well-established mechanisms to evaluate the regional CEA framework and outcomes. The ALCES® model should be subject to continuous refinement as more and better information becomes available. Model predictions should be tested and assumptions revisited.

Explicit incorporation of results into agency activities

The long-term development of the regional CEA will be predicated on its usefulness to participating agencies. The CMP should endeavour to use the modelling results in planning exercises as soon as possible in order to evaluate and promote the benefits of the process. The CMP should also strive to use the regional cumulative effects process to provide a better context for the evaluation of local development proposals.

Lessons learned

For the benefit of those pursuing similar projects, the authors have identified a number of key lessons that arose in the implementation of this project. These lessons can be categorized as: broad global lessons, lessons pertaining to the operation of such a project, lessons regarding cooperation, and lessons related to institutional frameworks.

Global lessons

Citizens are increasingly demanding that resource management agencies take an integrated approach to managing natural resources. The public is becoming more sophisticated in its understanding of resource issues, and greater conflict is ensuing as the footprint of human activity on the landscape expands. The values that citizens bring to the table are a critical part of the decision-making calculus and must be incorporated into policy, planning and management processes. Many natural resource management issues represent wicked problems and require the development of new and more interdisciplinary methodologies.

The development of cumulative effects models can contribute significantly to coping with wicked problems. The use of a cumulative effects model cannot drive a process, but needs to be seen as a tool that supports it. There is potential to put too much emphasis on the technical details of the model and lose sight of the overall purpose. The model should serve the process and not vice versa.

The indirect benefits arising from the collaborative process of addressing cumulative effects may actually outweigh specific project outputs and 'deliverables'. For example, the benefits that arise from promoting cooperation between agencies, the development of professional working relationships across jurisdictions, data sharing and the development of education materials are all valuable products of the exercise.

Operational lessons

In complex, multi-agency projects, having a 'lead agency' may create suspicion among the other participants. Having a neutral, third-party organization to facilitate the project is one way to overcome this issue, but requires that this entity has a clear mandate and adequate financial support. The development of a multi-agency steering committee with shared responsibility for chairing regular meetings is another mechanism to ensure that the process is truly collaborative. The value of strong leadership and 'champions' cannot be over-stated.

Complex integrated approaches to resource management must take a long-term view. Project operation must remain flexible to respond to such things as changing budgets, election timetables, personnel changes and workloads. Long-term commitment by participants is essential, as these complex processes offer no 'quick fixes'.

Participants should try to focus on problems that transcend boundaries and are common to the project membership. The development of common language and consistent data protocols is fundamental to long-term success. In large-scale collaborative approaches, there can never be too much communication. Finally, integration needs to be considered as a core activity by all participants and cannot be viewed as a collateral activity.

Lessons relating to institutional frameworks

Strategic approaches do not fit well with current political cycles and institutional structures, which emphasize short-term planning cycles. However, people 'on the ground' are more likely to recognize the importance of integration required for collaborative cumulative effects assessment. Collaborative processes to develop and implement integrated resource management may help to catalyse systematic changes to how wicked problems are addressed.

Conclusions

The Crown Managers' Partnership is an important initiative to promote collaborative transboundary approaches to ecosystem management in the Crown of the Continent. The partnership is currently comprised of government agency representatives and provides an effective forum for communication and information sharing. The interest in shared project development has resulted in an evolving framework for a regional cumulative environmental effects assessment. A strategic-level approach employing ALCES® is currently being evaluated and compared with other modelling approaches. The intent is to model 'what if' scenarios that focus on issues and resources of shared regional concern. Results of the process are not meant for direct application at the tactical or operational level, but are aimed at the development of higher-level regional strategies.

Individual representatives to the CMP have a shared understanding and commitment to understanding regional cumulative effects, although 'fine-tuning' of project goals and objectives is required. A variety of competing pressures have thwarted rapid development of the regional landscape assessment project, but the process has been valuable and will contribute to addressing the overall wicked problems that are inherent in large-scale cumulative effects.

Afterword

Subsequent to the first drafts of this chapter being prepared, the Crown Managers' Partnership decided not to pursue the RLAP process as described above. The undertaking was considered too onerous for the current level of support and capacity available to the group. Moreover, it was decided that additional information and strategic direction were required to effectively embrace such an exercise. However, the CMP remains committed to a regional collaboration and integrated management. To this end, the CMP has drafted a strategic plan to guide its efforts toward maintaining and enhancing ecological health in the Crown of the Continent Ecosystem. The strategic plan outlines an approach to improve understanding, raise awareness and promote collaboration around the topic of ecological health. Additionally, the CMP identified the need to strengthen and stabilize the capacity of the organization. The approach still requires a considerable effort around data/information collection, sharing and harmonization. As such information is compiled and assessed, it may provide an opportunity to revisit a more formal process of modelling and managing cumulative effects.

References

Baxter, W., Ross, W. A. and Spaling, H. (2001) 'Improving the practice of cumulative effects assessment in Canada', *Impact Assessment and Project Appraisal*, vol 19, pp253–262

Brown, A. L. and Therivel, R. (2000) 'Principles to guide the development of strategic environmental assessment methodology', *Impact Assessment and Project Appraisal*, vol 18, pp183–189

Cocklin, C., Parker, S. and Hay, J. (1992a) 'Notes on cumulative environmental change I: concepts and issues', *Journal of Environmental Management*, vol 35, pp31–49

Cocklin, C., Parker, S. and Hay, J. (1992b) 'Notes on cumulative environmental change II: A contribution to methodology', *Journal of Environmental Management*, vol 35, pp51–67

Contant, C. K. and Wiggins, L. L. (1991) 'Defining and analyzing cumulative environmental impacts', *Environmental Impact Assessment Review*, vol 11, pp297–309

Creasey, J. R. (2002) 'Moving from project-based cumulative effects assessment to regional environmental management', in Kennedy, A. (ed.) *Cumulative Environmental Management, Tools and Approaches*, Calgary, Alberta Society of Professional Biologists

Damman, D. C. (2002) 'The challenges of developing regional frameworks for cumulative effects assessments', in Kennedy, A. (ed.) *Cumulative Environmental Management, Tools and Approaches*, Calgary, Alberta Society of Professional Biologists

Dickens, P. (2003) 'Changing our environment, changing ourselves: Critical realism and transdisciplinary research', *Interdisciplinary Science Reviews*, vol 28, pp95–105

Drouin, C. and LeBlanc, P. (1994) 'The Canadian Environmental Assessment Act and cumulative environmental effects', in Kennedy, A. (ed.) *Cumulative Effects Assessment in Canada: From Concept to Practice*, Calgary, Alberta Society of Professional Biologists

Dube, M. G. (2003). 'Cumulative effect assessment in Canada: A regional framework for aquatic ecosystems', *Environmental Impact Assessment Review*, vol 23, pp723–745

Forem Technologies (2006) 'ALCES: An integrated landscape management tool', www.foremtech.com/products/pr_alces.htm accessed October 2006

Griffiths, A., McCoy, E., Green, J. and Hegmann, G. (1998) *Cumulative Effects Assessment*, Calgary, Macleod Institute

Hansen, A. J., Rasker, R., Maxwell, B., Rotella, J., Johnson, J. D., Wright, A., Parmenter, U. L., Cohen, W. B., Lawrence, R. L. and Kraska, M. P. V. (2002) 'Ecological causes and consequences of demographic change in the new west', *BioScience*, vol 52, pp151–162

Hegmann, G., Cocklin, C., Creasey, R., Dupuis, S., Kennedy, A., Kingsley, L., Ross, W., Spaling, H., Stalker, D. and AXYS Environmental Consulting Ltd (1999) *Cumulative Effects Assessment Practitioners Guide*, Hull, Quebec, AXYS Environmental Consulting and CEA Working Group for the Canadian Environmental Assessment Agency

Kahn, A. E. (1966) 'The tyranny of small decisions: Market failures, imperfections, and the limits of economics', *Kylos*, vol 19, pp23–47

Kennett, S. A. (1999) *Towards a New Paradigm for Cumulative Effects Management*, Calgary, Canadian Institute of Resource Law

Kreuter, M. W., DeRosa, C., Howze, E. H. and Baldwin, G. T. (2004) 'Understanding wicked problems: A key to advancing environmental health promotion', *Health Education and Behavior*, vol 31, pp441–454

Long, B. (2002) *Crown of the Continent: Profile of a Treasured Landscape*, Kalispell, MT, Crown of the Continent Ecosystem Education Consortium

Marsden, S. (2001) 'An international overview of strategic environmental assessment, with reference to World Heritage Areas globally and in Australian coastal zones', *Journal of Environmental Assessment Policy and Management*, vol 4, pp31–66

Nelson, H. G. and Stolterman, E. (2003) *The design way. Intentional Change in an Unpredictable World – Foundations and Fundamentals of Design Competence*, Englewood Cliffs, NJ, Educational Technology Publications, Inc.

Noble, B. F. (2002) 'The Canadian experience with SEA and sustainability', *Environmental Impact Assessment Review*, vol 22, pp3–16

Piper, J. M. (2002) 'CEA and sustainable development evidence from UK case studies', *Environmental Impact Assessment Review*, vol 22, pp17–36

Prato, T. (2003) 'Alleviating multiple threats to protected areas with adaptive ecosystem management: The case of Waterton-Glacier International Peace Park', *The George Wright Forum*, vol 20, pp41–52

Rittel, H. J. and Webber, M. M. (1973) 'Dilemmas in a general theory of planning', *Policy Science*, vol 4, pp155–169

Ross, W. A. (1998) 'Cumulative effects assessment: Learning from Canadian case studies', *Impact Assessment and Project Appraisal*, vol 16, pp267–276

Shoemaker, D. J. (1994) *Cumulative Environmental Assessment*, Department of Geography Publication Series No 42, Waterloo, University of Waterloo

Slocombe, D. S. (1994) 'Ecosystem integrity as a basis for cumulative effects assessment', in Kennedy, A. J. (ed.) *Cumulative Effects Assessment in Canada: From Concept to Practice*, Calgary, Alberta Society of Professional Biologists

Stinchcombe, K. and Gibson, R. (2001) 'Strategic environmental assessment as a means of pursuing sustainability: Ten advantages and ten challenges', *Journal of Environmental Assessment Policy and Management*, vol 3, pp343–372

Travis, W. R., Theobald, D. M. and Fagre, D. B. (2002) 'Transforming the Rockies: Human forces, settlement patterns, and ecosystem effects', in Baron, J. S. (ed.) *Rocky Mountain Futures: An Ecological Perspective*, Washington DC, Island Press

Turner, M. G. and Carpenter, S. R. (1999) 'Tips and traps in interdisciplinary research', *Ecosystems*, vol 2, pp275–276

Wear, D. M. (1999) 'Challenges to interdisciplinary discourse', *Ecosystems*, vol 2, pp299–301

11

The Muskwa-Kechika Management Area: The Failings of a Multidisciplinary Rather than an Integrated and Interdisciplinary Approach

Paul Mitchell-Banks

Brief description of project area

The Muskwa-Kechika Management Area (M-KMA) is a unique management area in northeastern British Columbia, Canada, encompassing more than 64,000km² of the Northern Rocky Mountains and Great Plains (see Figure 11.1). It is arguably one of the most significant wilderness areas in North America (M-KMA, undated), and possibly the largest unroaded wilderness south of 60°N on the continent. The M-KMA has the potential to remain one of the largest wilderness areas in North America, but there are many challenges in addressing interdisciplinary and integrated planning and management.

From March 2000 until March 2002, I served as the Muskwa-Kechika programme manager, the senior government manager responsible for management and planning in the area. This was a unique position within the Government of British Columbia, charged with championing the legislative intent of addressing the diametrically opposed issues of, on one hand, wilderness and wildlife management in a globally significant area and, on the other, simultaneously promoting and facilitating resource extraction from this area, which also has world-class oil, gas and other values.

While some attention was paid to a multidisciplinary approach intended to support a globally unique land-use planning and management process, the success of the effort was severely limited because it did not have an interdisciplinary and integrated approach. Such an approach would have increased coordination in the land-use planning, management and research processes that would have informed each other throughout their existence – rather than generating information, at the end of the project, that rarely effectively incorporated information and data from other processes.

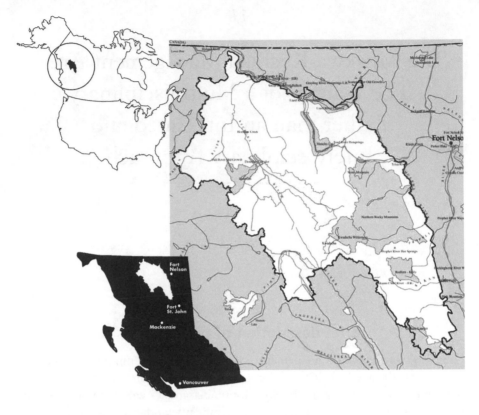

Figure 11.1 Location of the Muskwa-Kechika Management Area

Source: www.muskwa-kechika.com

Multidisciplinary research has often been assumed to effectively address the challenges of issues that cross over a number of areas. Two definitions are:

> *Multidisciplinary research projects are defined as those that involve two or more investigators from different disciplines that focus different perspectives and capabilities on complex problems that intersect established areas of study.* (Wichita State University, 2005)

> *Research that incorporates scientists or studies from a number of disciplines. Basically, it is research that includes the ideas and has studies in many different fields, for example in archaeoastronomy and paleoethnobotany. The idea of multidisciplinary research is really not restricted to a specific time period; rather, as long as people have participated in research they have included research from other fields in their studies.* (Wilson, 1997)

While these definitions sound promising, they fail to emphasize the critical need to integrate information gained from all disciplines throughout the process, rather than simply generating reports from a number of disciplines and attempting to integrate them together at the end with some summary and calling that 'multidisciplinary'. This failure to incorporate, integrate and effectively consider the multiple perspectives across disciplines throughout the entire process results in much multidisciplinary research falling short of its potential. At the same time, interdisciplinary research (IDR) also has its own challenges, which have to be effectively addressed to maximize its potential (Committee on Science, Engineering and Public Policy, 2004):

> *Despite the apparent benefits of IDR, researchers interested in pursuing it often face daunting obstacles and disincentives. Some of them take the form of personal communication or 'culture' barriers; others are related to the tradition in academic institutions of organizing research and teaching activities by discipline-based departments—a tradition that is commonly mirrored in funding organizations, professional societies, and journals.*

The project

The M-KMA was delineated by three approved land and resource management plans (LRMPs) that addressed the Fort Nelson, Fort St John, and Mackenzie areas of northeast British Columbia (Land Use Coordination Office, 1997a, b, 2000). LRMPs are multi-year, multi-stakeholder processes, mandated by government and with agency participation and support, that attempt to integrate land and resource management in a high-level plan that essentially creates a strategic 'road map' for future land-use planning and management by government ministries. Each of the three LRMPs proposed a central area of the three adjoining regions that should receive special management consideration. The M-KMA was formally designated by a 1997 Order-in-Council (through which cabinet ministers approve a decision without going to the legislature) with the creation of the Muskwa-Kechika Management Plan (Land Use Coordination Office, 1997c).

In 1998, the government passed the Muskwa-Kechika Management Area Act. This directly addressed the specifics of the management area and plan, the establishment of the Muskwa-Kechika Advisory Board (a public advisory group), the Muskwa-Kechika Trust Fund and general issues such as the power to make regulations. Both the management plan and the act require five additional planning processes. One innovative aspect of the legislation was the explicit attempt to incorporate multi-ministry sign-off on two of the more challenging plans, ensuring that planning was coordinated, addressing a broad range of issues and not necessarily just those of one ministry. The idea behind this multi-ministry approach was to ensure the incorporation of a number of disciplines and legislative perspectives in order to make the planning more comprehensive. The five plans and responsible ministries were as follows:

- Recreation Management Plan – Ministry of Environment, Lands and Parks; Ministry of Forests;
- Oil and Gas Pre-tenure Plans – Ministry of Energy and Mines; Ministry of Environment, Lands and Parks; Ministry of Forests;
- Landscape Unit Objectives – Ministry of Forests;
- Wildlife Management Plan – Ministry of Environment, Lands and Parks;
- Parks Management Plan – Ministry of Environment, Lands and Parks.

After the government responsible for establishing the M-KMA was defeated, the new government scrapped the multi-ministry philosophy, so that all the plans are now signed off by single ministries. The Ministry of Tourism, Sport and the Arts is now responsible for the Recreation Plan, and the Ministry of Energy, Mines and Petroleum Resources for the Oil and Gas Pre-tenure Plans.

The aim of both the M-KMA legislation and the management area is to directly address the perceived extraordinarily difficult task of maintaining the globally significant wilderness and wildlife values of the area while also promoting and facilitating economic development, with the oil and gas and potential mineral values taking a very high profile (M-KMA, undated). This project involved two sets of management challenges, most of which were not fully anticipated at the time the legislation was drawn up and the management area established: administrative and legal management challenges, and sectoral management challenges.

Administrative and legal management challenges

Access
The most obvious administrative and legal management challenge stems from the combination of the size, geography and limited access of the M-KMA. Its area is 6.4 million hectares, larger than European countries such as Switzerland, Belgium, The Netherlands, Slovakia, Austria and Estonia. There are very few roads, and those established are anticipated to be temporary, used for oil and gas and mineral exploration and development, and forestry operations. No additional roads or trails are being developed for recreation or general access.

Weather conditions create another management challenge, with seasonal fluctuations of 70°C not uncommon. At -40°C, equipment is subject to heavy wear and all staff are required to take wilderness survival courses and carry emergency survival gear (cold weather sleeping bags, first aid kits, rations, satellite phones and so on) in their vehicles.

The limited road access creates very real management challenges for government staff. Other means of travel are often employed, including horseback, helicopter, jetboat, snowmobile and all terrain vehicles (ATVs). Fuel often has to be remotely cached, due to the long distances from staging points and the size of the M-KMA. Access was one of the most contentious issues identified by the three LRMPs that led to the establishment of the M-KMA. Within the M-KMA, a separate planning process was conducted to tightly regulate access in the area through the access management plan. Its aim is to minimize the numbers of access routes into the M-KMA and thus, hopefully, to control the resulting access impacts.

Another management challenge is the long distance (approximately 1300km by road, two hours by air) between the regional managers and the senior government managers located in the provincial capital. There was an ongoing effort to keep both sides informed, with the regional (northern) managers more aware of the operational concerns, and the southern (provincial) managers more aware of the ongoing political environment and privy to more information about future legislative and policy initiatives. This sometimes led to challenges trying to match 'intent' with 'realistic management options'.

Planning and plans

One ongoing challenge is attempting to address the overall intent of three LRMPs in the planning for the M-KMA. The LRMPs tend to deal with generalities and do not give specific direction with respect to operational planning. This lack of specificity can create challenges when the regional manager has to evaluate and then make the inevitable trade-offs that planning entails. This challenge is further compounded by declining government budgets and staffing, making the attainment of objectives derived during earlier better-funded periods more difficult.

This lack of specific guidance is also complicated for the M-KMA with its legislated requirement of five additional plans, as noted above. The legislation only specifies the broad activities that the plans are supposed to address; it provides no guidance with regard to their scale and scope, detail and so forth, and does not specify whether any plan ranks higher in terms of legislation or policy than the others. This is compounded by the practical and legal uncertainty about what exactly constitutes 'wilderness'. This very hotly debated concept incorporates issues such as man-made impacts and the presence or absence of people, and is a key component of the M-KMA legislation. However, as discussed below, a proposed set of definitions was developed only recently, and these have yet to be accepted by the government (Muskwa-Kechika Advisory Board, 2005).

Land claims

The final, and potentially greatest, challenge revolves around the land claims of Native Indians (indigenous people) that cover virtually all of the Province of British Columbia. Treaty Eight was signed over a century ago, and three of the advisory board members represent bands (communities) that signed the treaty.

It is currently being reviewed by the provincial government due to the dissatisfaction of the Treaty Eight bands (Ackerman, personal communication, 10 March 2006). The Kaska Dena Treaty is currently under negotiation and four board members represent Kaska Dena bands. Any legislation, decisions or actions within the M-KMA are 'without prejudice' with respect to a future treaty with the Kaska Dena. A Letter of Understanding, signed in 1997 between the provincial government and the Kaska Dena Council (Government of British Columbia and Kaka Dena Council, 1997), permits land-use planning and other government management initiatives to continue pending the signing of a treaty. Thus, planning and management are under way without any future certainty with respect to what lands may be awarded to the Kaska Dena in the north and particularly in the M-KMA.

Sectoral management challenges

Oil and gas

The oil- and gas-rich Western Canadian Sedimentary Basin extends into the northeast corner of British Columbia, including the M-KMA. The oil and gas sector is currently the most active and most visible in the area, and will probably create the greatest management challenges. Increases in both oil and gas commodity prices over a number of years have led to greater interest in investing in exploration and development, and a sharp increase in the number of seismic, drilling, facilities and pipeline applications. While most oil and gas activity to date is outside the M-KMA boundaries, high prices lead to increasing pressure to more actively explore and potentially develop the area.

Pre-tenure planning has to be completed for the relevant Special Management Zone. These zones are areas of the M-KMA in which economic development is permitted as long as it is done in acceptable ways that address wilderness and wildlife values. Oil and gas applications have been consistently increasing in the recent past, and this is not expected to stop. In British Columbia, oil and gas is regulated by the Oil and Gas Commission. During the fiscal year ending 31 March 2005, 3067 new applications (wells, pipelines and geophysical) were received, representing a 9 per cent increase over the previous year, which was also a 9 per cent increase over the preceding year (2002–2003) (see Figure 11.2).

Oil and gas development requires roads to access well sites, pumping stations and, often, pipelines. While helicopters are sometimes used, this is not the usual practice due to the higher cost. Roads and seismic lines not only increase access to humans but are also used by predators such as wolves (James and Stuart-Smith, 2000; Mladenoff et al, 1995; University of Alberta, 2006), potentially increasing their ability to hunt over a larger area with suspected higher predation on moose, caribou and elk – three important species in the M-KMA.

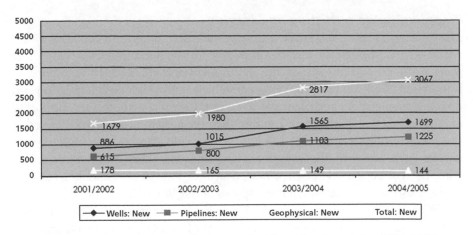

Figure 11.2 New oil and gas applications to the Oil and Gas Commission (2001–2005)

Source: Oil and Gas Commission (2004/2005)

Forestry

The M-KMA has limited commercial timber resources (Land Use Coordination Office, 1997c). There is long-term planning for potential limited forestry development, but economic harvesting is constrained by road development costs, log hauling distances, stand size and density and piece size. It is likely that forestry development will be closely associated with the oil and gas sector, taking advantage of roads approved for the oil and gas exploration and development and, potentially, any mineral exploration and development. The most promising areas are located at the south end of the M-KMA in the Graham/North Special Management Zone.

Mineral resources and energy

The M-KMA has potentially significant mineral resources. The government has funded a mineral occurrence database that supports existing tenure holders and exploration and development activity. While the area has received limited exploration (Land Use Coordination Office, 1997d), this has recently changed with soaring global commodity prices and increased interest in coal, copper and even diamond mining within the area (Ackerman, personal communication, 10 March 2006). As with forestry, one challenge to mining is the very limited and controlled access – but there are opportunities to explore how to develop mines in conjunction with other industrial activities such oil and gas and forestry.

Increasing energy prices and the growing desire to develop more sustainable power generation have led to the suggestion of locating large wind farms within the M-KMA. This is currently being discussed with the Muskwa-Kechika Advisory Board (Ackerman, personal communication, 10 March 2006).

Trapping, guide-outfitting and tourism

Historically, trapping and guide-outfitting have been important activities within the M-KMA. Though the number of people involved in this sector is far fewer than in the oil and gas sector, the government and the LRMPs recognize its importance. This has led to a strong commitment, through both legislation and ongoing management and planning, to ensure that these activities are supported and have their needs considered among all of the industrial activities. Related to this, recreationists, including hunters, fishers and the backcountry tourism sector, have accessed the area for many decades. This group is primarily comprised of both local and provincial residents, but national and international tourists also visit the vast wilderness spaces of the M-KMA.

With regard to tourism, the M-KMA is important for two distinct user groups: local residents who invest large sums of money and time to access the area by plane, horse, jetboat and ATV; and wealthy hunters (primarily European or American) who pay premium prices for guided hunts for Stone's sheep, grizzly bear, elk, moose and mountain goat. Despite this tourism activity, a number of ongoing issues need to be effectively addressed:

- The First Nations people who have resided in the area for thousands of years have virtually no involvement in tourism activities.
- Tourism and marginal economies – no systematic plan or strategy has been developed to determine how to best use tourism to address the marginal economies of many of the smaller communities in the area, particularly the First Nations communities.
- Limits of experience – there has been no systematic evaluation of the tourism potential and lessons learned/impacts noted to date, not only in this area, but in comparable areas.
- Acceptable limits of impacts – a conservation area design (CAD) (Groves, 2003) was completed in 2005, and is a conservation-focused 'biology-based toolkit' that can provide data critical to land use, resource planning and management (Muskwa-Kechika Advisory Board, 2006). Unfortunately, the CAD commenced well into the planning process and this delay prevented it from serving as a basis for the other five planning processes. There is also no ongoing funding for monitoring and mitigation – especially with respect to revising the CAD to incorporate new data, knowledge, technologies or developments.
- Limits and spatial marginalities – there is still little or no understanding of limits of acceptable change (Stankey et al, 1985) and dealing with spatial marginalities (cross-boundary issues, remote areas and the larger context within which an area exists).

These issues are further discussed in the penultimate section of this chapter.

Time-frame and funding sources

Muskwa-Kechika Trust Fund

In addition to the standard funding that goes to government ministries to carry out their work in northeast British Columbia, the Muskwa-Kechika Trust Fund provides additional funding that focuses on the specific challenges of the M-KMA. The fund was created by the M-KMA Act, and has two primary purposes:

1   to support wildlife and wilderness resources of the management area through research and integrated management of natural resource development;
2   to maintain in perpetuity the diversity and abundance of wildlife species and the ecosystems on which they depend throughout the management area.

The trust fund currently receives an annual base funding of C$2 million (US$1.75 million), as well as a project fund top-up allowance in which the government will match dollar-for-dollar contributions, to a maximum of C$500,000 (US$440,000) (Muskwa-Kechika Advisory Board, 2006). The base funding and the matching funding mechanism create a potential maximum annual fund of C$3 million (US$2.6 million). This fund has experienced a number of changes to its maximum funding level and funding mechanisms (composition of base and matching funding), and uncertainty about the long-term funding level has created a tremendous challenge to the advisory board and government managers in strategically funding projects.

Every autumn, there is a call for project proposals that are submitted to the advisory board, subjected to a number of review processes, and then reviewed by the board that passes recommendations to the trustee of the trust fund (the Minister of Agriculture and Lands; previously the Ministry of Sustainable Resource Management). After he has reviewed and approved the recommended projects, management of the projects is assigned to the Muskwa-Kechika programme manager, who serves as the trust fund projects manager and comptroller. Funding agreements addressing deliverables, reporting and funding arrangements are negotiated and signed. The five current project funding envelopes under the trust fund are:

1   building an information base;
2   supporting planning;
3   improving management;
4   advancing applied science (research);
5   promoting awareness and involvement.

Disciplines represented in the project

The government staff directly involved in activities within the M-KMA come from a number of disciplines, including forestry operations, biologists, conservation officers, oil and gas managers and British Columbia Land and Water officials involved in land and water tenures and licensing. Ideally, interdisciplinary concepts and working approaches are needed to address the very challenging issues faced in effectively planning and managing the M-KMA. There is an ongoing effort through the Peace Managers Committee (PMC) and the Interagency Management Committee (IAMC) (both addressed below in the section on stakeholders) to share information, but this is at a high level. Equally, it would be desirable to have staff exchanges and specialized staff brought in for complicated projects, but these options are constrained by staffing levels and funding. This lack of effort is driven by inadequate numbers of staff and very restricted funding. One example is that northeast British Columbia, with an area of over 200,000km^2, far larger than many European countries, has only six biologists in the government's regional office in Fort St John. The lack of resources is further complicated by the vast distances and challenging and expensive travel – expensive helicopters and jetboats are widely used to access the M-KMA.

Stakeholder involvement

Stakeholders are involved in a number of ways in this initiative, and play differing roles both in the definition of issues, priorities and activities and in the implementation of research. Five groups of stakeholders are involved in this process. The First Nations and economic (oil and gas, guide-outfitters and trappers, forestry, mining, wind power) stakeholders have been described above. Others, as described below, are the LRMP teams, the Muskwa-Kechika Advisory Board, and government staff.

Land and resource management planning teams
Each of the three LRMPs was created through extensive stakeholder participation with government staff and sent to the British Columbia government for review, potential amendment and, ultimately, approval. The planning teams have continued to play a role in the LRMP process since the formal approval of the LRMPs by the government, as they have annual compliance meetings to monitor how the government implements the formally approved LRMP. As the three LRMP areas overlap the M-KMA, all three LRMP teams have continued monitoring involvement.

Muskwa-Kechika Advisory Board
As noted above, the Muskwa-Kechika Advisory Board is appointed by the premier of the Province of British Columbia. Its roles include providing advice

on natural resource management in the M-KMA and identifying suitable projects and proposals to receive trust funding (Minister of Environment, Lands and Parks, 1998).

The initial advisory board had up to 17 members to represent a broad range of interests, including but not limited to: First Nations, environmental, business, labour, and representatives of the Fort Nelson and Fort St John LRMP teams (Clark, 1998). The current advisory board is slightly larger, including: the chair; seven First Nations representatives; three local government representatives; two members each from the oil and gas sector and from environmental interests; one representative from each of the mining, forestry and backcountry tourism sectors, labour, trapping/guide-outfitters, and the British Columbia Wildlife Federation; and the Muskwa-Kechika programme manager (ex-officio position). The board membership was expanded to include a representative of backcountry tourism and a second oil and gas representative, to reflect the importance of those sectors.

While the advisory board has no statutory power, its members have a very strong ability to lobby government and other stakeholders, and play a key role in the initiative, as they:

- provide advice and recommendations to the government about management and planning within the M-KMA;
- provide recommendations on expenditures from the trust fund (which have historically been accepted);
- participate in various planning processes, as well as LRMP monitoring teams;
- can serve as a 'sounding board' for government staff to raise issues and seek advice.

Government staff

The government is arguably the key stakeholder in the management of the M-KMA, as it has the legislative authority and ultimate responsibility. Government staff attempt to ensure that statutory requirements (defined by legislation) are upheld and administered. They also play a key role in different settings:

- *Their relationship with the advisory board.* This is a key relationship given the role of the board and the need to minimize conflict and have a coordinated approach to the land-use planning and management of the area. Good relations with the advisory board can lead to additional funding from the trust fund being directed to initiate or support projects proposed by government staff that have direct relevance to the M-KMA. When conflict arises, members of the board can make things more difficult for government managers, given their ability not only to speak with authority about the area when raising issues, but also to directly lobby politicians and high-level bureaucrats to bring attention to government staff and/or actions that they are not pleased with.

- *Their ministry role in northeast British Columbia.* The M-KMA is only part of the provincial area for which the various government ministries are responsible. The allocated resources (staff, facilities and budget) are always under pressure, and this can lead to difficult decisions in terms of where to focus time and efforts. The government staff are constantly challenged to meet their ministry responsibilities throughout 'their' entire region, and this challenge is increased by the fact that the planning and management standards within the M-KMA often exceed those required outside of the area.
- *Their ministry role in the PMC.* The PMC was established by the regional managers to assist in information sharing and coordination between the ministries in northeast British Columbia, an area that incorporates the Peace River and is referred to as 'The Peace'. Collectively, the PMC has a more effective voice in government and with the advisory board than the individual ministries.
- *Their ministry role in the IAMC.* The IAMC represents the senior government managers from all the ministries involved in land-use planning and management within northern British Columbia. The M-KMA programme manager serves on the IAMC in a unique role, as he has no statutory power and represents a specific area of the province with a very high profile.

Implications of the project

Local/regional management decisions

The failure to conduct a CAD process at an early enough stage and to subsequently implement the five legislated planning processes in a logical fashion has led to a planning effort that is not only out of sequence but also inadequately informed. Though conservation concerns were the core concept behind establishing the area, they are not optimally addressed for wilderness; wildlife; economic and industrial activities, such as oil and gas, forestry and mining; or tourism activities, such as guide-outfitting, recreational hunting, wildlife viewing, hiking and camping.

A CAD of the M-KMA was completed in 2005, but this was severely constrained by a number of factors, including but not limited to:

- limited attention being paid to boundary areas and adjacent issues;
- the late commencement of the CAD in the planning process, after some pre-tenure planning had been completed, the Parks Management Plan had largely been completed, and the Recreation Management Plan had been completed for only the original two-thirds of the area – as it did not include the area from the Mackenzie LRMP that was added to the M-KMA later as that LRMP was completed some years after the Fort Nelson and Fort St John LRMPs (Land Use Coordination Office, 2000);

- the speed of the process, with a view to obtaining as much value as possible with the tight time pressure created by ongoing planning efforts.

An ideal process to rectify these problems would be to immediately halt any new industrial and economic activity within the M-KMA until the completion of the following, specifically in this order:

- The Wildlife Plan, which focuses on maintaining intact ecosystems and animal populations to ensure ecological integrity. The draft plan now offers two levels of guidance: one to government decision-makers and one to industry. The first part provides a results-based description of what is expected by all in the M-KMA, while the second, more detailed, part provides specific guidance to wildlife managers and offers some excellent background to decision-makers and industry.
- The Recreation Management Plan that encompasses the entire M-KMA, as well as addressing boundary issues and adjacency concerns.
- Parks management plans for all parks, rather than the management direction statements that apply to most of the parks in the M-KMA. These are typically about a dozen pages long and insufficiently detailed for effective planning. They provide a general description of what the park has been set aside for, with a less detailed description/prescription on its uses. Budget restrictions have led to more of these being utilized. While they provide government staff with a general direction to be followed and work very well for small parks, they have the disadvantage of leading to inconsistent scale and scope for management and planning.
- Landscape unit objectives planning to address landscape planning and effective, ecologically sustainable and economically achievable forestry management, planning and operations.
- Pre-tenure planning to ensure ecologically sound and secondarily economically achievable oil and gas management, planning and operations.
- Finally, while completing the five legislated plans, a tourism management plan would be developed. This would inform the other planning processes, and also ensure that all economic and industrial activities consider each other, and that specific trade-offs are made to ensure that all activities are accounted for effectively and none are addressed as an 'afterthought'.

This is a rather idyllic and theoretical approach and, given the lack of coordinated planning to date, it is politically and economically unacceptable to put a halt to oil and gas operations and tourism activities; forestry is limited at this time. However, this lack of forethought and commitment to sound planning and management leads to impacts on wilderness, wildlife and ecosystems and a series of conflicts between the economic sectors. Consequently, a more realistic approach would be:

- Announce a moratorium on any new economic and industrial activities until the planning processes are completed.
- Limit oil and gas activities to the less contentious areas where oil and gas pre-tenure planning has been completed. This is a difficult issue to address effectively, given the importance of oil and gas development and its strong economic return to the government. The Upper Sikanni Pre-tenure Plan was the first plan, completed within the resources and timeframes available. It has been reviewed, resulting in a report on its updating to bring it more in line with the other plans. Subsequent plans cover the Besa-Prophet, Muskwa-West, and Sulphur/8-mile Special Management Zones.
- For a time-limited period, allocate additional funding and resources to ensure timely and effective completion of planning. This would ensure that processes involving the government, the advisory board and public participation were not only completed but, more importantly, completed in a rational sense. Furthermore, the time-limited effort would also reduce the opportunity for various parties to promote projects that do not contribute meaningfully to the entire M-KMA initiative. This has been attempted at various times by the provincial government, First Nations, the advisory board and corporate interests in the M-KMA.
- Come to a land claims settlement, not only with the Kaska Dena Nation but also addressing the outstanding issues that remain 100 years after the Treaty Eight settlement with a number of First Nations with interests in the M-KMA.

This approach would require the provincial government to reverse a number of policy decisions made by the existing government and by its predecessor. These policy decisions have severely curtailed the success of the M-KMA initiative and, in some cases, effectively sabotaged it. One example is the arguably preferential treatment given to the oil and gas sector, which has impacted both the forestry and tourism sectors, and undermined integrated planning and management in the M-KMA.

Local/regional policies

The existence of the M-KMA and the related research, planning and management are taken into consideration in local community planning and management. The regional communities view the M-KMA as a valuable resource to promote local economic development. Participation by community residents in both the LRMPs and the advisory board serve to keep these policies on the table when land-use planning and management concerns are discussed.

National policies

The M-KMA has national significance within Canada. It is difficult to assess its impacts on national policies, but there is evidence that it does influence policy-making; it is one of the case studies used by the National Round Table on the

Environment and the Economy, a high-profile policy group (R. McManus Consulting Ltd. and Salmo Consulting Inc., 2004).

Other research/projects

The trust fund has played a key role in promoting other research and projects, with C\$900,000 (US\$790,000) having been given to the University of Northern British Columbia and Northern Land Use Institute to fund a Muskwa-Kechika professor and to support research (Province of British Columbia, 1999). Research funded by the trust fund is public property and accessible to researchers, with some limitations with regard to sensitive data, such as locations related to spiritual sites and critical wildlife areas. Such sensitive data can be accessed through confidentiality agreements, and in the case of First Nations work, there would probably have to be some additional agreement reached with the relevant First Nations.

Lessons for integrated research and management

There have been a number of principal challenges to integrated management and research in the M-KMA initiative. This section addresses the lessons learned and potential strategies to overcome some of the shortcomings of the process to date. The eight major lessons learned to date concern:

1 the importance of definitions and common understanding;
2 adequate data and knowledge;
3 accurate and timely data analysis and decisions;
4 adequate funding;
5 politics and agendas;
6 roles and responsibilities;
7 tourism, the forgotten or ignored sector;
8 the need to create a formal intent and culture to promote interdisciplinary and integrated planning.

The importance of definitions and common understanding

Definitions and common understanding can play a key role in IDR and integrated management and planning. Definitions act as anchors or focal points for initiatives or projects; they need to be specific, with a common understanding from all members of the planning, management and research teams. Four key concepts have not been adequately defined or addressed in the M-KMA: wilderness, carrying capacity, limits of acceptable change, and cumulative impacts. Furthermore, the final three concepts are not referred to in the M-KMA legislation, and as such there are no formal regulatory processes in place to address them.

Wilderness

Wilderness is one of the two key values whose maintenance is a goal of the legislation establishing the M-KMA, drawn up in 1998. However, no accepted definition of the word 'wilderness' yet applies to the area. This is an issue for both the advisory board and government staff and, in fiscal year 2004–2005, the advisory board established a project to clearly define the terms 'wilderness', 'wilderness quality' and ' wilderness characteristics'. The aim was to ensure that management of activities occurs in a manner consistent with the intent of the M-KMA Act. The advisory board submitted its wilderness definition to the provincial government in 2005, intending that this should be tested by managers and utilized as a policy tool as well as at an operational level (Muskwa-Kechika Advisory Board, 2005). This definition still awaits a formal response from the government, and until there is an accepted definition, there is effectively no common understanding of what constitutes wilderness.

The lack of a clear definition has been a significant flaw in the planning effort. Planning requires a process that is traceable, accountable and reproducible (Mitchell-Banks, 1999), and without a clear definition of wilderness this is virtually impossible to achieve. Traceable not only refers to a planning process that is not only documented but involves specific objectives, values, considerations and issues that are addressed. Accountable refers to who is responsible for what; without a definition of wilderness, the accountability of any government department/individual manager for this resource remains uncertain. Reproducible refers to being able to retrace or re-enact the planning effort, as logical concerns and strategies are documented and continue to be valid if conditions have not changed.

The most widely discussed definition of wilderness is that found in the US Wilderness Act, 1964, which has globally served as a key conceptual piece or framework of legislation. Wilderness, as defined in this law, has three equally important characteristics:

- it is not to be controlled by humans, natural ecosystem processes operate freely and its primeval character and influence are retained;
- it is not occupied or modified by mankind, humans are merely visitors and the imprint of their work is hardly noticeable;
- it offers outstanding opportunities for solitude or for a primitive and unconfined recreation experience (Society of American Foresters, 1989).

This definition is inappropriate for the M-KMA, where there is a deliberate effort to maintain wilderness and wildlife values while permitting economic and industrial development to various degrees in both the Protected Areas and the Special Management Zones. Nevertheless, this classical wilderness definition can serve as a starting point and can be modified to address the unique nature of the intent and planning of the M-KMA.

Carrying capacity

The concept of carrying capacity was initiated by the conceptual work of Wagar (1964). This concept proposes that an ecosystem has a limited ability to absorb and recover from impacts before becoming degraded. Hardin (1991) describes carrying capacity as 'the fundamental basis for demographic accounting'. Conventional economists and planners have not accorded this concept much credence when it is applied to human beings (Rees, 1996). Each ecosystem is considered to have a specific carrying capacity for the population of each plant or animal species, and this can fluctuate from year to year depending on climate and other factors (Chiras, 1988). The very nature of the M-KMA initiative, namely to maintain the wildlife and wilderness values while permitting economic development, demands not only a good understanding of carrying capacity but also effective means to measure and monitor its states. Failing to do this effectively implies a process that is not being managed against any performance level or standard, which makes it very difficult, if not impossible, to achieve truly sustainable management.

Limits of acceptable change

The concept of limits of acceptable change (LAC) was first proposed by Frissell (1963) and developed further by Frissell and Stankey (1972). It has two dominant management objectives (Cole and Stankey, 1998) of staying within maximum acceptable deviations from both the natural range of variation in ecological conditions, and a pristine wilderness experience. Stankey et al (1985) published a framework for establishing acceptable and appropriate resource and social conditions in recreation settings. This has played a major role in the research and practice of LAC and, over time, the application of LAC has been extended beyond wilderness, into the front country of national parks, and even into addressing issues like the impacts of wood-burning stoves (Cole and McCool, 1998). However, LAC cannot be used in all situations, certain key aspects of the land-use planning, management and research must be applicable (Cole and McCool, 1998):

- if there is no conflict between goals, there is no need for an LAC process;
- if there is conflict between goals, but one goal cannot be compromised, an LAC process is not appropriate;
- if managers cannot establish a hierarchy of goals, some of which will serve as constraints, LAC will not work;
- even for management issues for which there is conflict and agreed room for compromise using a hierarchy of goals, the LAC process can only be applied if it is possible to establish measurable and attainable standards that quantify the minimally acceptable state of the goal that serves as the ultimate constraint.

The M-KMA has two primary goals – the maintenance of wilderness and wildlife values – and these should serve as the constraining goals. Failure to establish baseline levels and to have in place measurable and attainable standards is fatal when establishing a useful LAC process. The Oil and Gas Commission has established geophysical guidelines for the M-KMA (Oil and Gas Commission, 2004), but without LAC conditions being met, they are arguably of very limited value.

Cumulative impacts

Cumulative impacts remain one of the most challenging of all aspects of land-use planning and management to successfully address. According to the Council on Environmental Quality (1978):

> *'Cumulative impact' is the impact on the environment which results from the incremental impact of the action when added to other past, present, and reasonably foreseeable future actions regardless of what agency (Federal or non-Federal) or person undertakes such other actions. Cumulative impacts can result from individually minor but collectively significant actions taking place over a period of time.*

This US legislative definition clearly refers to incrementality over all time spans and the potential for collective impact on the ecosystem. The difficulty is in measuring and monitoring cumulative impacts, and to this end the baseline status of the ecosystem has to be established, followed by constant monitoring for change. This is far harder to implement than just specifying an environmental concern in legislation. Measurement of the impacts of an economic or industrial activity within the M-KMA must be carried out in a scientific and statistically valid manner to manage cumulative impacts and avoid overwhelming the area's carrying capacity. Only by doing this continually is it possible to support all economic sectors, both resource extraction and tourism. Even for tourism it is necessary to manage visitor use – especially when it is desirable to maximize tourist revenues (Henry, 1992).

While the Muskwa-Kechika Trust Fund did fund a cumulative impact report (Muskwa-Kechika Advisory Board, 2005), this was fairly general, with limited direct application to the area, as it did not establish baseline environmental standards or refer to specific tools and methods for the M-KMA. The whole question of cumulative impacts is being looked at by the government (Ackerman, personal communication, 10 March 2006) and in the M-KMA there are ongoing research efforts, such as the five-year project entitled 'The Ecosystem Approach to Habitat Capability Modelling and Cumulative Effects Management', to further develop an approach to addressing cumulative impacts (Muskwa-Kechika Advisory Board, 2005). An immediate effort to establish a continual means of formally monitoring the cumulative impacts within the M-KMA is required. Its primary intent would be to ensure that

wilderness and wildlife values are protected as a first priority with any industrial and economic activities being subservient to those two values.

Adequate data and knowledge

The second lesson revolves around the challenges of having or obtaining
adequate data and knowledge about the wilderness qualities, wildlife and habitat
values, natural resource values and land management challenges of the M-KMA.
The difficulties stem from not only the area's spatial scale but also its tremendous
biodiversity, differences in physio-geography, and problems of access.

The LRMPs were completed under a government that was not as time- or
cost-conscious as the current government. This has led to challenges when the
current government cut back on staffing and budgets; the same challenges
remain with fewer resources to address them. A lesson here is to formally
commit (though legislation) to adequately following up on unique land planning and management initiatives to ensure their successful implementation.
Another strategy is to aggressively use triage and a project management
approach to identify the critical steps in the process that are time-sensitive and
to focus on those and practise risk management with respect to the budget and
timeline. In the case of the M-KMA, this requires cooperation between the
advisory board and the government that was not always in place, as the parties
did not agree on priorities. In retrospect, the government should have
mandated that all planning be completed prior to the trust fund being used for
any other approved purposes, such as training, extension or non-critical
research (Mitchell-Banks, 2003).

Accurate and timely analysis and decisions

The third challenge is related to the previous lesson, but centres around the
challenges of accurate and timely analysis and decision-making. This challenge
involves a number of components, including translating field data into a
geographic information system (GIS), and devising adequate tools to verify,
correct, analyse, manipulate and map the output. There is ongoing interest in
how GIS could be used to undertake sensitivity analysis and scenario planning
to permit planners and management teams to have a better understanding of
potential consequences or outcomes.

There are also the challenges of timing with respect to planning requirements, implementation and ongoing operational management. Politicians and
industry can push a timeline to finish planning and create 'certainty' that leads
to challenges at best, and failure at worst, to complete planning in a
comprehensive and integrated fashion. One set of challenges concerns out-of-
date inventories and, even worse, missing inventories. Integrated and
interactive planning tools, such as CAD, have been completed on timelines
that do not inform planning from the beginning. It has been necessary to

attempt to creatively fit tools into the truncated planning timelines – which severely limits the benefits and values that can be obtained from the tool development and application processes. Ideally a CAD would have been completed before any resource use was contemplated. This is legislatively possible within the M-KMA but, due to heavy government and industrial pressures, planning has been accelerated, and planners and land managers make the best with what they have and can obtain within the planning timelines. Similarly, incomplete and/or out-of-date inventories can severely constrain the accuracy or value of management decisions and planning. The adage 'garbage in, garbage out' applies to planning as much as it does to models and simulations (Mitchell-Banks, 2003).

Related to the previous point is the question of effective policy. A process to evaluate the short-, medium- and long-term implications of planning actions or the lack thereof is needed. It is tempting for the government to think short-term and focus on revenue generation, particularly when there are fiscal challenges. Short-term solutions can result in inadequate planning, leading to financial losses through less than satisfactory resource planning (suboptimal planning for effective and economic use or extraction) or greater costs (associated with environmental restoration and remediation). As important, and arguably ethically more so, is the potential for species and habitat losses. There is a strong argument that these should be fully incorporated into the evaluation, development and implementation/administration of any policy.

Adequate funding

The fourth challenge involves obtaining adequate funding for travel, staff and equipment to effectively manage the area. Related to this is the necessity to be able not only to address ongoing management needs, but also to proactively address anticipated developments – such as oil and gas exploration and development continuing to increase in both scale and scope. Hiring freezes and challenges in finding and retaining staff (particularly with GIS expertise) compound this problem. The trust fund provides some security in potential funding, but all projects proposed by the government for trust funding have to be vetted by the advisory board, which makes recommendations to the trustee before his approval. This has led to some frustration in the advisory board over the difference between government core responsibilities (which the trust fund is not supposed to give money for) and activities incremental to base funding. The frustration is mirrored by government staff, who 'see' the money being potentially available in the trust fund, but not necessarily accessible for government activities (Mitchell-Banks, 2003).

When drawing up the initial legislation or policy to establish the area, real attention should have been paid to the future implications and the costs, timelines and most effective interdisciplinary and integrated approach to the planning and management required. Equally, a fund-raising strategy should have been initiated as soon as the area was established. Fund raising is only now beginning to be seriously looked at; some very positive media opportunities while the initiative was still new have been lost. However, the current challenge can be addressed by means that include:

- committing to finishing one initiative or related initiatives prior to taking on new challenges; and avoiding the temptation to spread resources thinly, attempting to do a little of everything;
- ensuring a clear understanding, through legislation and administrative procedure, on how funds and staff time can be employed; and being consistent with the approach while being flexible to address real emergencies and not just funding 'nice to do' activities.

Politics and agendas

Politics and agendas are the fifth challenge, and are an inherent part of the M-KMA initiative. There are politics in the provincial capital, with politicians responding to lobbying and making decisions with inadequate (or at times no) consultation with their own staff – often with very frustrating consequences, as decisions can be made with no appreciation of the existing situation or resources available to government staff responsible for the area. The advisory board is very political; every member has a constituency and an agenda to address within the area. Government managers and staff are also political in that they have the mandate of their ministries to fulfil, and the conflicting agendas of ministries can lead to clashes. One of the most challenging aspects of the programme manager's job is to champion the M-KMA initiative and serve as a liaison between the various groups and promote effective communication, 'buying' into a common vision and cooperation. This job plays a keystone role in not only holding the process together but also encouraging it to progressively continue forward (Mitchell-Banks, 2003). This position could play a key role in the promotion and support of IDR.

Roles and responsibilities

Roles and responsibilities are the sixth major lesson. The advisory board is unique in the province not only for the scope of its advisory and monitoring function, but also because it makes recommendations for a multi-million dollar trust fund. In many ways, it is the trust fund that can potentially serve the greatest role in ensuring the success of the initiative.

One requirement for the trust fund to be effective is a healthy appreciation between the advisory board and government staff in how to best apply it to complement government statutory responsibilities and initiatives. The trust fund is there to advance the M-KMA with a view to maintaining the unique features of the area while also permitting industrial and economic development over three-quarters of the area. More interdisciplinary research could be funded by the trust fund to more effectively address the outstanding land-use planning and management challenges. There needs to be a greater accountability for decisions and a more effective effort to inform the general public about how the funds are being spent (Mitchell-Banks, 2003).

Tourism, the forgotten or ignored sector

The M-KMA tourism sector mainly comprises two distinct tourism groups: first, resident and non-resident hunting, often served by transporters and guide-outfitters; and second, river rafting, photo tours, fishing, hiking, horseback tours and so on.

Despite the intent of the legislation and the creation of the management board and trust fund, the tourism sector has not received the same attention as other sectors (Mitchell-Banks, 2006). This may change with the Ministry of Tourism having taken over recreation and the recreation planning for the M-KMA. Dual-track tourism, one of the fastest growing and lucrative tourism streams, incorporates wilderness/wildlife activities with cultural learning and offers one of the most promising options (Central Coast Consulting, 2002). Tourism has been overshadowed by the lucrative oil and gas sector and has played a minor role in terms of planning and management.

The need to create a formal intent and culture to promote interdisciplinary and integrated planning

Interdisciplinary and integrated planning is a difficult process (Committee on Science, Engineering, and Public Policy, 2004):

> *Individual researchers involved in interdisciplinary research (IDR) require a supportive environment that permits them to work in multiple disciplines and departments and to be fairly evaluated and rewarded for both their interdisciplinary and their disciplinary work. They have a responsibility to explain and demonstrate the benefits of IDR, venture into new fields, and be open to the cultures and values of other disciplines... Researchers need opportunities to train in two or more disciplines and to work closely with faculty members and students in each. Such cultural and intellectual immersion is a prerequisite to high-quality interdisciplinary work. Researchers may need to spend considerable time on activities (teaching, research, committees, and community service) outside their home department.*

This quote captures some of the challenges for IDR for academics, and the challenges for government researchers are often greater, as there is usually less flexibility in terms of time and resources and flexibility between ministries. It is not common for government staff to gain experience in other ministries in order to broaden their appreciation of other disciplines or perspectives – a critical aspect of IDR. Having an ongoing formal programme of government researcher exchange would greatly facilitate IDR work.

Bringing in outside experts also helps, as do workshops and conferences, but these are often 'after the fact' and not incorporated into project design and coordination with government staff. A failing of the trust fund allocations was not to emphasize the primary need to focus on IDR and integrated work, resulting in a collection of projects with some overlap but not enough inter-disciplinary and integrated focus. The IAMC and PMC only partially serve to bring a broader perspective to this, and when I served as the Muskwa-Kechika programme manager, my multidisciplinary background helped to some degree, but I supervised the research contracts funded by the M-KMA trust fund and not the research itself.

Summary

The M-KMA is arguably one of the most globally significant wilderness, wildlife and economic extraction areas in the northern hemisphere. The values are world-class, but there are very significant planning, management and research challenges, compounded by limited time, resources and staff. An attempt has been made to make this a multidisciplinary process, but the constraints of resources and industry and political pressure to expedite the process have all served to severely constrain the initiative.

The M-KMA is full of challenges, successes and, unfortunately, some failures. The challenges include: large spatial scale, access, climate, the objective of maintaining wilderness and wildlife values while permitting economic development, pressure from the oil and gas sector to expedite planning and thus access to the area, limited government staff and resources, and a lack of formal processes and ability to manage carrying capacity, LAC and cumulative impacts.

Successes include: the designation of the area as a special management region, the Access Management Plan, and the establishment of parks and protected areas (which do allow for limited low-impact development) and Special Management Zones (for economic development). The advisory board and the trust fund are also key successes, as is the generally close cooperation between government staff and the advisory board.

The planning is complex and has consumed a tremendous amount of effort and resources, and staff and stakeholders have generally entered into the processes with a generous and cooperative attitude. However, the plans completed to date have not achieved their potential. The three major contributing factors are:

1   a failure of the legislation for the area to be clear about the relationship of the plans to each other, the order in which they should be completed and their order of precedence;
2   a lack of time, finances and resources (primarily staff) on the government side and very limited funding to contract out work that should have been completed to best contribute to the planning and management – despite some truly heroic efforts on the part of the staff;
3   missing or delayed planning processes that would have been powerful tools to assist in planning, management and research. These processes would have explicitly addressed issues such as the definitions of wilderness and acceptable economic development, and the establishment of a formal means to evaluate cumulative impacts, carrying capacity, and determining and monitoring LAC.

This shortfall is best exemplified by the accelerated pre-tenure planning that was completed in some areas before the CAD and wildlife management plans. The lack of a formal plan for maintaining wilderness values reduces the profile of the challenge and, in some ways, contributes to it not always being front and centre in all research, planning and management.

A well-planned, adequately funded interdisciplinary and integrated planning, research and management process would have led to greater success – particularly if there had been a stronger understanding of the challenges and opportunities at the beginning of the process, and a political will to 'stand fast' and resist the pressure of industry. Such strong resilience on the part of the government and the advisory board would have helped ensure that all planning and management activities were adequately informed by research and monitoring.

Despite these criticisms, the process is globally significant and, while flawed in a number of ways, represents a tremendous effort on the part of the advisory board and government staff. The initiative continues and it is hoped that additional lessons will be incorporated into the planning, management and research in a timely and effective manner.

References

Ackerman, A. (2006) Regional Manager, Peace Region, Ministry of the Environment, Province of British Columbia, personal communication, 10 March 2006
Central Coast Consulting (2002) *Nicola Tribal Association Eco-tourism Study*, Merritt, Central Coast Consulting
Chiras, D. D. (1988) *Environmental Science, A Framework for Decision Making*, Don Mills, Benjamin/Cummings Publishing Company Inc.
Clark, G. (1998) *Muskwa-Kechika Advisory Board Terms of Reference*, Victoria, Government of British Columbia
Cole, D. N. and McCool, S. F. (1998) 'Limits of acceptable change and natural resources planning: When is LAC useful, and when is it not?' in McCool, S. F. and Cole, D. N. (eds) *Proceedings – Limits of Acceptable Change and Related Planning Processes: Progress and Future Directions*, General Technical Report INT-GTR-

371, Ogden, US Department of Agriculture, Forest Service, Rocky Mountain Research Station

Cole, D. N. and Stankey, G. H. (1998) 'Historical development of limits of acceptable change: Conceptual clarifications and possible extensions', in McCool, S. F. and Cole, D. N. (eds) *Proceedings – Limits of Acceptable Change and Related Planning Processes: Progress and Future Directions*, General Technical Report INT-GTR-371, Ogden, US Department of Agriculture, Forest Service, Rocky Mountain Research Station

Committee on Science, Engineering, and Public Policy (2004) *Facilitating Interdisciplinary Research*, Washington DC, The National Academies Press

Council on Environmental Quality (1978) *Regulations for Implementing NEPA*, http://ceq.eh.doe.gov/nepa/regs/ceq/1508.htm#1508.7, accessed November 2006

Frissell, S. S. (1963) *Recreational Use of Campsites in the Quetico-Superior Canoe Country*, St Paul, University of Minnesota

Frissell, S. S. and Stankey, R. G. (1972) 'Wilderness environmental quality: Search for social and ecological harmony', *Proceedings of the 1972 National Convention, Hot Springs, 1–5 October*, Washington DC, Society of American Foresters, pp170–183

Government of British Columbia and Kaska Dena Council (1997) *Letter of Understanding*, Victoria, Government of British Columbia

Groves, C. R. (2003) *Drafting a Conservation Blueprint: A Practitioner's Guide to Planning for Biodiversity*, Washington DC, Island Press

Hardin, G. (1991) 'Paramount positions in ecological economics', in Constanza, R. (ed.) *Ecological Economics: The Science and Management of Sustainability*, New York, Columbia University

Henry, W. R. (1992) *Carrying Capacity, Ecological Impacts and Visitor Attitudes: Applying Research to Park Planning and Management*, Paper presented at the Kenya Ecotourism Workshop, Kenya, Washington DC, National Parks Service

James, A. R. C. and Stuart-Smith, A. K. (2000) 'Distribution of caribou and wolves in relation to linear corridors' *Journal of Wildlife Management*, vol 64, pp154–159

Land Use Coordination Office (1997a) *Fort St John Land and Resource Management Plan*, Victoria, Government of British Columbia

Land Use Coordination Office (1997b) *Fort Nelson Land and Resource Management Plan*, Victoria, Government of British Columbia

Land Use Coordination Office (1997c) *Oil and Gas Exploration and Development in the Muskwa-Kechika*, Victoria, Government of British Columbia

Land Use Coordination Office (1997d) *Mineral Exploration and Mine Development in the Muskwa-Kechika*, Victoria, Government of British Columbia

Land Use Coordination Office (1997e) *Muskwa-Kechika Management Plan*, Victoria, Government of British Columbia

Land Use Coordination Office (2000) *Mackenzie Land and Resource Management Plan*, Victoria, Government of British Columbia

Minister of Environment, Lands and Parks (1998) *Bill 37-1998. Muskwa-Kechika Management Area Act*, Victoria, Government of British Columbia

Mitchell-Banks, P. J. (1999) *Tenure Reform for Facilitating Community Forestry in British Columbia*, unpublished PhD thesis, Vancouver, University of British Columbia

Mitchell-Banks, P. J. (2003) 'Protecting and sustaining wilderness values in the Muskwa-Kechika management area', in Watson, A. and Sproull, J. (eds) *Science and Stewardship to Sustain Wilderness Values: Seventh World Wilderness Congress Symposium*, Ogden, US Department of Agriculture, Forest Service, Rocky Mountain Research Station

Mitchell-Banks, P. J. (2006) 'The Muskwa-Kechika Management Area: The failed planning and management of the Serengeti of the north', in Ryan, C. (ed.) *Taking Tourism to the Limits*, Singapore, Elsevier

Mladenoff, D. J., Sickley, T. A., Haight, R. G. and Wydeven, A. P. (1995) 'A regional landscape analysis and prediction of favorable gray wolf habitat in the northern Great Lakes region', *Conservation Biology*, vol 9, pp279–294

Muskwa-Kechika Advisory Board (2005) *Muskwa-Kechika Advisory Board 2004-2005 Report to the Premier and the Public*, Fort St John, Muskwa-Kechika Advisory Board

Muskwa-Kechika Advisory Board (2006) www.muskwa-kechika.com/trustfund accessed November 2006

M-KMA (Muskwa-Kechika Management Area) (undated) 'Homepage', http://ilmb www.gov.bc.ca/ilmb/lup/lrmp/northern/mk/index.html accessed November 2006

Oil and Gas Commission (2004) *Geophysical Guidelines for the Muskwa-Kechika Management Area*, Fort St John, Oil and Gas Commission

Oil and Gas Commission (2004/2005) *Annual Service Plan Report*, Fort St John, Oil and Gas Commission

Province of British Columbia (1999) 'Province funds Muskwa-Kechika Research Program at UNBC', http://ilmbwww.gov.bc.ca/lup/lrmp/northern/mk/news_releases/nr04999.htm accessed May 2007

R. McManus Consulting Ltd and Salmo Consulting Inc. (2004) *Muskwa-Kechika Case Study*, prepared for the National Round Table on the Environment and Economy

Rees, W. E. (1996) 'Revisiting carrying capacity: Area-based indicators of sustainability', *Population and Environment: A Journal of Interdisciplinary Studies*, vol 17, pp195–215

Society of American Foresters (1989) *Report of the Society of American Foresters' Wilderness Management Task Force*, Bethesda, Society of American Foresters

Stankey, G. H., Cole, D. N., Lucas, R. C., Petersen, M. E. and Frissell, S. S. (1985) *The Limits of Acceptable Change (LAC) System of Wilderness Planning*, General Technical Report INT-176, Washington DC, United States Department of Agriculture, Forest Service

University of Alberta (2006) www.biology.ualberta.ca/faculty/stan_boutin/ilm/index.php?Page=959 accessed November 2006

Wagar, J. A. (1964) *The Carrying Capacity of Wild Lands for Recreation*, Forest Science Monograph 7, Washington DC, Society of American Foresters

Wichita State University (2005) 'News release from the Office of Research Administration', www.niar.twsu.edu/oraweb/Forms/MURPA%20Guidelines.pdf accessed November 2006

Wilson, S. M. (1997) 'Introduction to Archaeology', www.utexas.edu/courses/wilson/ant304/glossary/glossary.html accessed November 2006

12

Ecological Restoration in the Canadian Rocky Mountains: Developing and Implementing the 1997 Banff National Park Management Plan

Clifford A. White and William Fisher

Banff National Park: Overview and management history

Introduction

A little over a century ago, the cultures of hunter-gathers that had occupied the Canadian Rockies for over ten millennia were rapidly replaced by an industrial culture advancing from both the Pacific and Atlantic Oceans. The government of Canada established large areas of national parks yet, within a few decades, long-term ecosystem states and processes were dramatically altered. The Bow Valley of Banff National Park (BNP) (see Figure 12.1) is the birthplace of Canada's national park system, established in 1885. It is also the busiest area of any Canadian national park, and one of the most developed landscapes (see Figure 12.2) within any national park in the world (BBVS, 1996). Due to historical development patterns, the Bow Valley holds a transcontinental railroad, the four-lane Trans Canada Highway, two large towns (Banff and Canmore), and another very busy visitor service centre (Lake Louise). Several million people per year pass through the BNP on the highway, and many utilize the national park trails and roads.

Yet, possibly because it is adjacent to large areas with lower development, the BNP is still utilized by a full assemblage of large native mammals with the exception of bison (*Bison bison*). Key indicator species for Rocky Mountain montane ecosystems include humans, wolf (*Canis lupus*), grizzly bear (*Ursus arctos*), black bear (*U. americanus*), elk (*Cervus elaphus*), bison, beaver (*Castor canadensis*), trembling aspen (*Populus tremuloides*) and willow (*Salix* spp.). Processes include human-caused wildlife mortality and displacement, human- and lightning-caused fire, predation and herbivory (see Figure 12.2).

Figure 12.1 Banff townsite area of the Bow Valley, Banff National Park
in 1902 (*top*) and 1997 (*bottom*)

Note: Riparian areas in the early photograph show willow and white spruce communities. Uplands sites are dominated by grasslands, shrub lodgepole pine, aspen thickets and isolated Douglas fir that resulted from frequent fires. The recent photograph illustrates the effects of fire suppression, highway construction and the urban development of Banff townsite, which has constricted wildlife movement to narrow corridors.

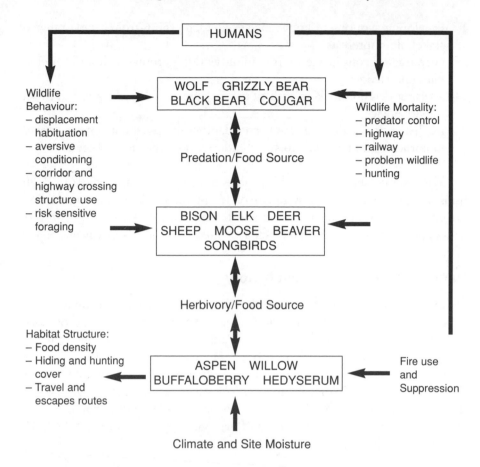

Figure 12.2 Montane ecosystem indicator states and processes for Banff Bow Valley

Source: Parks Canada (2002)

Management of this suite of species and processes has been difficult in BNP (BBVS, 1996; Hebblewhite et al, 2005) and particularly controversial in Yellowstone National Park (Huff and Varley, 1999; Ripple et al, 2001), Jackson Hole, Wyoming (Clark, 1999; Hessl, 2002) and Rocky Mountain National Park, Colorado (Hess, 1993; Baker et al, 1997) in the western US. Of great concern is the decline in abundance of palatable browse species, such as willow and aspen, due to high ungulate herbivory levels when combined with the restoration of the long-term regime of relatively frequent fires (White et al, 1998).

In this chapter, we describe landscape-level management in BNP, focusing on the Bow Valley. We first provide a brief synopsis of evolving ecosystem and national park management paradigms. Next we describe a recent ecosystem restoration programme that involved four broad components:

1   developing working or knowledge evaluation groups from a wide-range of stakeholder interests;
2   encouraging consilience by providing working groups with multidisciplinary knowledge;
3   advising decision-makers with potential outcomes based upon interdisciplinary advice;
4   implementing actions through an adaptive management approach with monitoring and feedback to stakeholders, scientists and managers.

We describe application of this process to ecosystem management in Banff, using as an example the restoration of long-term predator/prey interactions disrupted by modern human land use in the montane ecosystem. We conclude with opportunities for future interdisciplinary research and adaptive management.

National park management history

National park management in North America has ranged from policies that encouraged intensive terrestrial and aquatic ecosystem husbandry of select species for direct human benefit (for example, facility protection from fire, wildlife viewing and fishing), to more ecological science-based policies that evolved from a natural regulation objective that minimized any human intervention, to a more biologically focused objective, termed ecological integrity. Each of these policies has been based on prevailing scientific paradigms for trophic interactions in ecosystems, and has been applied to biological conservation in the BNP.

Tourism development and resource husbandry (1880s–1960)

Most North American national parks were established to develop tourism for economic benefit (Bella, 1987). In the case of Banff, the federal government initially kept most lands under public ownership to remove unregulated development around the hot springs, while working with the Canadian Pacific Railway company to develop tourism activities for visitors to the company's Banff Springs Hotel (Hart, 2003). The elimination of forest fires was important to protect forests, buildings, stock fences and bridges. Timber harvesting was encouraged in many areas to provide fuel breaks from fires and a local source of building materials and mine props (Nelson and Byrne, 1966). Park wardens killed predators such as wolves, cougars (*Felis concolor*) and coyotes (*Canis latrans*) (Jones, 2002) to increase populations of ungulate species most easily visible to tourists such as elk, bighorn sheep (*Ovis canadensis*), white-tail deer (*Odocoileus virginianus*) and mule deer (*Odocoileus hemionas*). From 1944 to 1969, domestic stock grazing management guidelines were utilized to determine culling levels for BNP's elk herd. Additional extensive development for tourism during this period included high-speed highways and mechanized ski areas (BBVS, 1996).

Natural regulation (1960s–1980s)

By the 1950s, the ecological roles of predation, herbivory and fire were increasingly appreciated, and scientists recommended adoption of 'natural regulation' policies. Predator control ended in Yellowstone National Park by 1950 and in BNP by 1965 (Jones, 2002). Elk culling stopped in both parks by 1970. Similarly, progress towards letting lightning fires burn began in the national parks of the western US in the 1970s (Pyne, 1995). A 'bottom-up' hypothesis for ecosystem function was the main scientific paradigm underlying natural regulation policy (Kay, 1998). Key tenets of natural regulation included:

- Predators generally remove only the weak and the infirm individuals (doomed surplus), thus in areas such as Yellowstone and the BNP where wolves were rare or absent, they could be regarded as 'non-essential adjuncts' (Cole, 1971).
- Vegetation and climate regulate ungulate populations at ecological carrying capacity (Caughley, 1979) where herbivore impacts may be relatively high in contrast to levels specified by range management guidelines.
- Humans were perceived to have a relatively insignificant long-term role in ecosystem function, particularly related to fire frequency (Johnson et al, 1995; Pyne, 1995), or hunting effects on populations of ungulates (Huff and Varley, 1999).

Ecological research on natural (for example, non-human) systems dominated. In the absence of cultural understanding, observed large-scale changes, such as declining in-park fire frequencies or wildlife population change, were often attributed to non-human factors such as climate change (Johnson and Larsen, 1991).

Ecological integrity and the long-term range of variability (after 1990)

A current primary mandate of national parks in Canada is to maintain or restore ecological integrity. Parks Canada Agency (2000) states that:

> *an ecosystem has integrity when it is deemed characteristic for its natural region, including the composition and abundance of native species and biological communities, rates of change, and supporting processes.*

A useful approach to evaluating ecological integrity is the historical or long-term range of variability (Landres et al, 1999; Morgan et al, 1994). The principles of the paradigm are: first, current ecosystems are the product of past conditions and processes; second, spatial and temporal variability in disturbance regimes are a vital attribute of ecosystems; and third, maintenance or restoration of long-term ecosystem states and processes will conserve biodiversity (Landres et al, 1999). The approach requires interdisciplinary scientific research to test predictions for the long-term condition of indicators defining ecosystem states and processes (for example, Kay et al, 1999). In many landscapes, careful, unbiased assessment of humans' past effects will be important (Kay and White, 1995). Humans may have been an important long-term ecological factor. For this reason, herein we avoid using the phrase 'natural range of variability' (Landres et al, 1999) because this describes a preconception of long-term conditions that may not be applicable to many ecosystems.

Some key questions related to this hypothesis include: how do top-down processes, such as predation and herbivory, interact with bottom-up processes such as climate change and fire? Can large carnivores be restored in modern

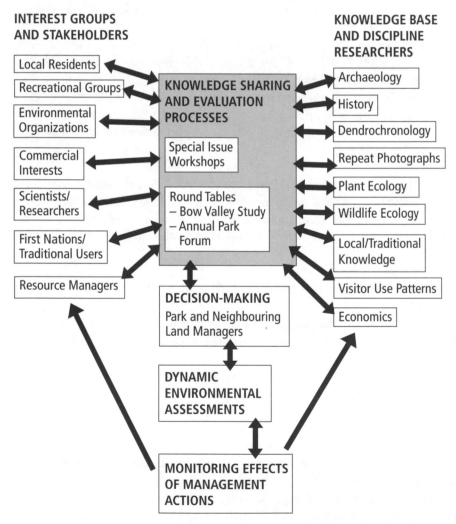

Figure 12.3 A model of collaborative and decision-making processes that guided ecosystem restoration in the Banff Bow Valley (1994–2004)

landscapes to population levels where the long-term predation process is functional? To what degree do modern human influences, such as highway-caused wildlife mortality and habitat fragmentation, impact the ecosystem? Where or when should historic human influences such as hunting and burning be restored? Where restoration actions such as highway mitigation, corridor restoration or prescribed burning appear necessary, how should they be prioritized and implemented?

Collaboration and interdisciplinarity in park planning

The paradigms for national park management described above were dominated by a current individual discipline or objective. Early park development was driven by economic development, resource husbandry by forestry and range management, and natural regulation by ideas then current in wildlife management. The concepts of ecological integrity and ecosystem management encourage broader citizen participation (Cortner and Moote, 1999). But how can this be facilitated when academic research is increasingly specialized and interest groups are increasingly polarized? Reviews of the BNP experience (for example, Draper, 2000; Hodgins et al, 2000; Jamal and Getz, 1999; Jamal et al, 2002; Wight, 2001; Zinkan and Syme, 1997) suggest a multiphase process (see Figure 12.3). First, it is essential to develop collaborative processes with stakeholders with a broad range of interests. Second, these groups need to be supported with a wide range of knowledge. Third, these groups need to provide decision-makers with recommendations based upon consilience developed through a collaborative social learning process. Fourth, actions initiated by managers should be implemented adaptively with feedback to both researchers and stakeholders. We describe each of these phases in more detail below.

Stakeholder collaboration and a future vision

As Canada's first and most developed national park, BNP has the mixed blessing of both great affection from many Canadians, but also great economic dependence on the tourism industry. Further, because the park includes a town and is close to the city of Calgary, local knowledge is abundant. These factors encourage strong citizen participation from many sectors of society (see Figure 12.3). Although Parks Canada has a long history of public participation and education, one of the strongest initiatives occurred during the Banff Bow Valley Study (BBVS) from 1994 to 1996 (BBVS, 1996). The minister responsible for Parks Canada mandated (BBVS, 1996) that:

> The Banff Bow Valley Study will be a comprehensive analysis of the state of the Bow Valley watershed in Banff National Park. The study will provide a baseline for understanding the implications for

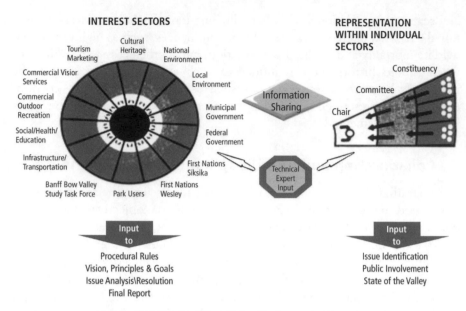

Figure 12.4 The Banff Bow Valley Study round table process

Source: BBVS (1996)

> existing and future development and human use, and the impact of
> such on heritage resources. The study will integrate environmental,
> social, and economic considerations in order to develop manage-
> ment and land use strategies that are sustainable and meet the
> objectives of the National Parks Act.

The core working group for the BBVS was an interdisciplinary task force and
secretariat (BBVS, 1996). The study's first objective was to develop a vision and
goals for the valley, which was achieved by assembling a round table (see
Figure 12.4) representing a broad range of interests, and informed by information from many disciplines (Hodgins et al, 2000; Ritchie, 1999). Ultimately,
the vision statement developed by the round table guided direction for the
overall park management plan, as stated by Parks Canada (1997):

> Banff National Park reveals the majesty and wildness of the Rocky
> Mountains. It is a symbol of Canada, a place of great beauty, where
> nature is able to flourish and evolve. People from around the world
> participate in the life of the park, finding inspiration, enjoyment,
> livelihoods, and understanding. Through their wisdom and fore-
> sight in protecting this small part of the planet, Canadians
> demonstrate leadership in forging healthy relationships between
> people and nature. Banff National Park is above all else a place of
> wonder, where the richness of life is respected and celebrated.

The practice of involving interest groups to develop and refine goals, and to advise park managers, continued after the BBVS with a round table discussion of park management directions and progress held at an annual park management forum. Further, several groups of stakeholders, experienced in the process of collaboration, continued to participate in shared decision-making processes, including the Banff Townsite Elk Advisory Committee, the Montane Science Advisory Group and the Lands Adjacent to the Town of Banff Advisory Committee.

Multidisciplinary knowledge

Quality information is required to feed a stakeholder review process. Although ideally it would be best if knowledge could be integrated and synthesized prior to presentation to stakeholders (that is, interdisciplinarity), in practice most research is quite specialized and is analysed and peer-reviewed under the paradigms prevalent in the individual disciplines (Endter-Wada et al, 1998). Thus, information available to stakeholders is most frequently provided in parallel streams, best termed 'multidisciplinary' (Balsiger, 2004). For the BNP example, we describe several disciplines that provided the most relevant information for the montane ecosystem (see Figure 12.3).

Archaeology

After the Pleistocene glacial recession (c.11,000 years BP), the area was immediately occupied by humans (Fedje et al, 1995). Faunal remains from archaeological sites on the east slope of the Canadian Rockies (summarized by Kay and White, 1995; Kay et al, 1999) have the following relative composition of number of individual bone specimens: bison (47 per cent), bighorn sheep (39 per cent), deer (7 per cent), elk (7 per cent), moose (*Alces alces*, <1 per cent) and goat (*Oreamnos oreamnos*, <1 per cent).

History

First-person explorer journal accounts from the Rockies for the period 1792–1872 (Kay et al, 1999, 2000) report generally low wildlife abundance. The relative number of recorded species hunted in the Rocky Mountains area was: bighorn sheep (54 per cent), bison (16 per cent), moose (12 per cent), goat (8 per cent), elk (5 per cent), deer (3 per cent) and caribou (*Rangifer tarandus*, 2 per cent). Kay et al (1999, 2000) attribute these long-term relative abundances to the effects of predation refuges. Bighorn sheep maintained populations due to refuge on steep slopes and cliffs. Bison were probably maintained by periodic immigration from the large Great Plains population to the east, which migrated hundreds of kilometres to minimize predation rates from wolves, humans and other predators (White et al, 2001). Elk apparently had few refuges from predation, and were maintained in long-term low abundance by humans and other predators (Kay et al,

Figure 12.5 Wolf and elk telemetry points around Banff townsite (1997–1999)

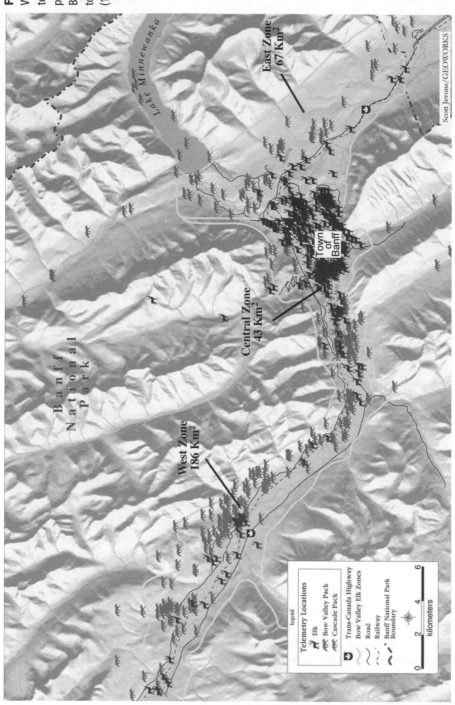

1999, 2000). Broad cultural changes related to European settlement resulted in further increases in human hunting rates of wildlife. In the Banff-Jasper area of the Rocky Mountains, the last bison sign was observed in 1859, and elk were reported as extremely rare by 1900 (Woods, 1990). In 1918 and 1920 a total of 235 elk from Yellowstone National Park were translocated to Banff (Woods et al, 1996).

Fire frequency research

Tree-ring studies indicate that long-term fire intervals in the Canadian Rockies may have been relatively short (<50 years) with low intensities on low-elevation valley bottoms (Tande, 1979; White, 1985). Fires may have been larger, less frequent and had higher intensity (flame length) and severity (depth of burn) at middle and higher elevation forests (Hawkes, 1980; Johnson and Larsen, 1991). This pattern of fire frequency, intensity and severity maintained a diversity of vegetation patterns (Rhemtulla et al, 2002; White et al, 2005) including savannah-like forests (see Figure 12.1, top) in the montane eco-region (Tande, 1979), and a patch-dynamic system in the subalpine (Johnson and Larsen, 1991). On the east slope of the Canadian Rocky Mountains, historic fires appear to have mostly been started by First Nations people, and later by early settlers and railroads. Evidence for the potential past significance of human ignition on the east slopes includes analysis of fire causes for the period 1880–1980 (White, 1985) and the low occurrence of lightning and lightning-caused fires on the east slopes compared to the west slopes (Wierzchowski et al, 2002). Further, dendrochronology studies show that past fire scars are mostly located in the dormant and early-wood sections of lodgepole pine (*Pinus contorta*) tree rings, not in the late wood (White et al, 2001). This suggests that most historic burning prior to 1880 occurred before or after the peak period of midsummer lightning occurrence, and was thus probably human-caused.

Wildlife ecology

Intensive research on wildlife occurred in conjunction with highway construction and after the Bow Valley of BNP was recolonized by wolves in 1985 (Paquet et al, 1996; Woods, 1990). Highway fencing reduced human-caused wildlife mortality (Clevenger et al, 2001, 2002; Woods, 1990) but due to wolf predation elk populations declined substantially in areas >5km from Banff townsite (Hebblewhite et al, 2002, 2005). However, near the town, intensive human development use and narrow wildlife corridors restricted carnivore movements (Duke et al, 2001), reducing predation rates in this area. By 1997, over 400 elk were concentrated in and around the town (McKenzie, 2001), surrounded by a 'halo' of wolves and other predators (see Figure 12.5). Long-term research on these trends indicated that natural regulation hypotheses that predators did not significantly influence Rocky Mountain ungulate abundance could be in error (Hebblewhite et al, 2005). Moreover, modern human use patterns in national parks could dramatically alter the predation process when unhunted elk become habituated to high human use levels, and thus obtain refuge from wary predators (White et al, 1998).

Researchers also intensively investigated grizzly bear population dynamics and habitat use in BNP and adjacent areas (Gibeau et al, 1996). These studies demonstrated a dramatic decline in the number of human-caused bear mortalities after food waste storage was improved (Benn and Herrero, 2002). Grizzly bear densities are highest in areas that have relatively low human use and include recently burned habitats (Gibeau et al, 2002; Hamer, 1996, 1999). In the 1990s, grizzly bear use was relatively low in the Bow Valley near the Trans Canada Highway (Chruszcz et al, 2003; Gibeau et al, 2002) except for near Lake Louise, where bears were frequently observed on ski runs cleared through mature forest (Gibeau et al, 1996).

Vegetation ecology

Repeat photography indicated that, historically, aspen and willow communities were lightly browsed and frequently burned (White et al, 2004), browse levels are now much higher, and conifer cover has increased dramatically (see Figure 12.1). Browsing studies show that, where elk densities were high (>5 elk per km²), herbivory removed all tall willows, with negative effects on beaver and some songbird species (Hebblewhite et al, 2005; Nietvelt, 2001). Even at only moderate elk densities (>2 elk per km²), herbivory levels were high enough to prevent aspen growth to >1m in height (White et al, 2003).

Human dimensions

Recognition that not just the footprint of development, but high levels of human use in general, could influence ecosystem processes such as predation led researchers to detailed computer mapping of human use levels (BBVS, 1996; Komex International, 1995). Detailed social science research developed understanding of use patterns including the timing of use, motivations for visitation and types of facilities used, and the social effects of forest and fire management programmes (for example, BBVS, 1996; McFarlane et al, 2004; Mountain Parks Visitor Survey Partnership, 2004). This knowledge was important to formulate feasible management alternatives. The local, regional and national economic benefits of the tourism and transportation industries in Banff are massive. Detailed research and modelling occurs periodically to quantify general regional patterns and trends (for example, Alberta Economic Development and Tourism, 1994; Cornwell and Cotanza, 1996). Commercial stakeholders are acutely aware of finer time and spatial scales of visitor use levels and expenditure patterns.

Collaboration and knowledge synthesis

The BBVS (1996) began the process of integrating knowledge by using the round table (see Figure 12.4) to critique a 'State of the Valley' compendium of information on ecological, economic and social systems (Pacas et al, 1996). Over a period of 10 months, the round table worked through various drafts to obtain a final acceptance of this report by all sectors (Hodgins et al, 2000). The BBVS task force, and to a limited extent the round table, guided an 'ecological outlooks project' (Green et al, 1996), which integrated several knowledge disciplines to evaluate potential outcomes of future development scenarios in the BBVS.

Each of the many disciplines described above provided some tentative evidence that Rocky Mountain montane ecosystems probably developed under a regime of long-term low human use levels, with associated anthropogenic hunting and burning patterns (Kay et al, 1999). Thus current high human use levels, with displaced predators, ungulates habituated to humans due to no hunting, and minimal anthropogenic burning, were a radical change from the long-term range of variability (White et al, 1998). However, the weight of evidence within each field of study on its own was insufficient to convince monodisciplinary researchers or peer reviewers to make this conclusion. In the absence of this integration, park management had largely continued to be guided by a mix of traditional economic development and natural regulation paradigms, assuming that development levels were not yet at a crisis point, predators were 'non-essential' adjuncts in montane ecosystem function, and that eventually an escaped lightning fire would rejuvenate vegetation.

The societal will to interpret and act on new evidence occurred when broadly based collaborative groups (for example, the BBVS round table and Banff Elk Advisory Committee) reviewed the multidisciplinary results and concluded that wildlife distribution and herbivory patterns around the town of Banff (see Figure 12.5) showed that the long-term range of variability was seriously disrupted by a major change in current versus long-term human land-use patterns (for example, from few hunter-gatherers to high densities of ecotourists). Ultimately, it was likely that the threat by aggressive elk to human safety triggered action. Local residents, who were dealing daily with the risks of human-habituated elk by carrying hockey sticks and slingshots, provided significant input to this stakeholder consensus! The Banff Bow Valley Study round table and task force, and subsequent special issue stakeholder groups used this common understanding to provide managers with numerous recommendations for restoring ecological integrity in montane ecosystems (BBVS, 1996).

Outcomes: Adaptive implementation of the 1997 Park Management Plan

The 1997 BNP Management Plan was a progressive response to the BBVS (1996) recommendations (Draper, 2000; Hodgins et al, 2000). The plan provided guidance for several restoration programmes described below. Recognizing the complexity of these projects, screenings required by the

Canadian Environmental Assessment Act were approved by park managers with the provision that:

- intensive monitoring programmes were in place;
- results of these programmes would be reviewed annually in collaborative processes including science workshops, advisory group committee meetings and the annual park planning forum;
- park managers adapted annual work plans for projects based on monitoring results and stakeholder review.

Highway mitigation

The high-speed Trans Canada Highway bisects the Bow Valley (see Figure 12.1b). Traffic volumes near Banff townsite increased from an average of about 8000 vehicles/day in 1982 to nearly 15,000 vehicles/day in 1994. Upgrading to divided four-lane standard was completed for 45km in BNP between 1979 and 1998, and required a comprehensive wildlife mitigation programme including fencing and 24 wildlife crossing structures (Clevenger and Waltho, 2000; Clevenger et al, 2001, 2002; Forman et al, 2003; McGuire and Morrall, 2000). Initially, the structures were mainly designed to facilitate ungulate crossings. However, stakeholder consultations for the Phase 3A (km 27–45) environmental assessment recommended greater attention to large carnivore crossings, and this, in combination with recommendations from the BBVS task force, led to construction of two large wildlife overpasses across the highway in this area. Subsequent monitoring has shown that these overpasses, in combination with wide underpasses (>50m) best facilitate highway crossings by wolves and grizzly bears (Clevenger et al, 2002).

Wolf recolonization

Wolves were eliminated from the southern Canadian Rockies by 1914, but recolonized BNP during a period of high ungulate abundance in the late 1930s (Cowan, 1947). A regional carnivore control programme again eliminated wolves from BNP in the 1950s, and consistent pack activity did not occur in the Bow Valley for nearly 30 years (Holroyd and Van Tighem, 1983). In 1985, the valley was again recolonized, probably due to dispersal from the nearby Red Deer Valley (Paquet et al, 1996). Management actions to assist recolonization included closures near den site areas, temporary restricted speed zones on roads through wolf activity centres, highway mitigation and wildlife corridor restoration. Wolf abundance and distribution was an icon indicator for stakeholders in many collaboration processes (Jones, 2002). Issues requiring resolution included the proposed fencing of the Banff townsite area to reduce elk avoidance of wolves (BBVS, 1996), supported by pro-wolf interests, versus no fencing with elk relocation, supported by town residents.

Human use management and wildlife corridor restoration

Human interference with wildlife movement around the town of Banff has been a long-term concern (BBVS, 1996). Environmental mitigation for a new housing subdivision on the edge of Banff townsite included permanent closure to human use of a wildlife movement corridor around the south perimeter of the town on Sulphur Mountain in 1997 (Golder and Associates, 1994). The BNP Management Plan (Parks Canada, 1997) required relocation of facilities including government and community stables, a small airstrip and a military cadet training camp out of the wildlife movement corridor around the north perimeter of the town below Cascade Mountain (Duke et al, 2001). East of the townsite, wildlife crossings over hydropower canals and penstocks were restored, and portions of the Minnewanka Road were closed in the winter. Additional human-use guidelines implemented after 1997 to minimize disturbance to large carnivores included mountain-biking restrictions on Bryant Creek, winter private vehicle use closure of the golf course, and extension of a summer closure of the middle Spray Valley (Parks Canada, 1997, 1999). Parks Canada worked with neighbouring land-management agencies to develop wildlife-corridor management guidelines for the whole Bow Valley (Bow Corridor Ecosystem Advisory Group, 1998, 1999). The Lands Adjacent to the Town of Banff Advisory Group continues to adapt and refine trail use in this area.

Human-use management issues remain controversial (Cooper et al, 2002; Petersen, 2000) regarding the reduction of human use on some trails to levels (<100 groups per month) recommended for grizzly bear security areas (Gibeau et al, 1996, 2001). Long-term collaboration between stakeholders, land managers and researchers (both social and ecological) will be required to develop consensus on potential management actions in order to conserve grizzly bears (Herrero et al, 2001). The 2003 amendments to the 1997 BNP Management Plan continue to refine principles for human use management and a revised decision-making framework for the conservation of grizzly bears.

Elk management strategy

The high concentration of elk near Banff townsite (see Figure 12.5) became a serious public safety concern in the 1990s (McKenzie, 2001) and also caused high herbivory impacts on montane willow and aspen communities (Hebblewhite et al, 2005; Nietvelt, 2001; White et al, 1998). The BNP Management Plan (Parks Canada, 1997) provided direction to 'restore predator-prey relationships' and 'restore vegetation communities to reflect the long-term ecosystem states and processes'. As described above, restoration of wildlife corridors increased predation rates near the town. In addition, Parks Canada relocated 217 highly human-habituated elk out of the park in 1998, 1999 and 2000 to reduce elk–human conflicts and herbivory impacts (Parks Canada, 1999). After 2001, an aggressive aversive conditioning programme with herding dogs was used to move elk out of the townsite area (Kloppers et al, 2005). The project was guided by an annual meeting of the Montane Science Workshop consisting of both scientists and stakeholders. Recommendations from these sessions were forwarded to the Elk Management Advisory Committee, and from this group on to the park superintendent. The Bow Valley elk population declined from >800 (1988), to <500 by 1998 to <200 by 2002. By 1998, willow growth began to exceed 1m in height throughout most the area (Hebblewhite et al, 2005; Nietvelt, 2001), and by 2004 aspen heights in many areas exceeded 1m (White et al, 2004).

Fire, forest thinning and mountain pine beetle control

The 1997 Park Management Plan required that 50 per cent of the long-term burn area be maintained: approximately 14km²/year (Eagles, 2002). As a result, the existing burning programme was expanded, and by 2004 a total of over 170km² had been treated in the previous 20 years (Pyne, 2004; White et al, 2005). The general procedure for burning is to use hand or mechanical thinning to build fuel breaks where required, followed by ground or aerial ignition during periods of low fire intensity to blacken unit boundaries and, ultimately, aerial ignition of the main units during warmer and drier conditions. Prescribed burning in BNP has been done almost totally in May and September, outside the peak period of high intensity burning conditions that usually occurs in July and August (Wierzchowski et al, 2002). The timing of 'shoulder season' burns thus likely partially corresponds to the long-term, anthropogenic pattern described above (White et al, 2005). After 2002, BNP rescheduled its burning programme and conducted limited tree removals along the east boundary of the park to reduce colonization levels by mountain pine beetles (Parks Canada, 2002). Prescribed burning, cutting of beetle-attacked trees and forest thinning options are presented annually to stakeholders and scientists at the Montane Science Workshop. Recommendations from this group are provided to the Park superintendent to adapt the programme.

Banff heritage tourism strategy

Implementing the above active restoration programmes required a high level of stakeholder and general public support. The tourism industry, following the recommendations of the BBVS (1996) established the Banff Heritage Tourism Corporation with members including the Banff/Lake Louise Tourism Bureau, Town of Banff, Town of Canmore, Parks Canada, the Banff Centre (a specialized educational institution) and the Banff Lake Louise Hotel Motel Association (Wight, 2001). The corporation developed a strategy with core objectives of:

- making all visitors and residents aware that they are in a national park and world heritage site;
- encouraging opportunities, products and services consistent with heritage and environmental values;
- encouraging environmental stewardship initiatives;
- strengthening employee orientation, training and accreditation programming for sharing heritage understanding with visitors.

The programme keeps Banff's business community involved, active and aware of ongoing ecological restoration efforts, annually trains hundreds of front line staff, and awards staff and businesses demonstrating environmental stewardship (Banff Heritage Tourism Corporation, 2004).

Communications

Possibly the most important component of BNP's restoration programme is communications to stakeholders and the public. Ongoing initiatives include a 'Research Updates' series presented each spring, topic information on the Parks Canada website, presentations at numerous meetings of stakeholders and NGOs, routine reporting through local, regional and national media, and outreach programming to local schools (Parks Canada, 1997).

National and regional implications

Banff is Canada's first and busiest national park, thus new management directions here often have important implications for other national parks and protected areas. The BBVS (1996) recommendations stimulated action at a national level for new management direction. In 1998, the Canadian Minster of Heritage established the 'Ecological Integrity Panel' (Parks Canada Agency, 2000). Similar to the BBVS, this interdisciplinary task force used a series of multi-stakeholder meetings to develop a future vision and broad set of recommendations for how Canadian national parks could maintain or restore ecological integrity.

Current Parks Canada national park management guidelines now also require developing a 'state-of-the-park' document every five years. These documents should be based upon ecosystem models (for example, Figure 12.2) developed in collaboration with stakeholders and researchers. The models will be used to define a standard set of indicators for biodiversity, terrestrial and aquatic ecosystems, and human dimensions. Ecosystem models, indicators and monitoring protocols will be developed at bioregional level (for example, Rocky Mountains) to provide a standardized approach and efficiencies for monitoring and research.

Regionally, the 1997 BNP Management Plan became a model for revising the plans for adjacent Jasper, Yoho and Kootenay national parks. The background state-of-park reports for these plans all use similar ecosystem models to

BNP (see Figure 12.2). Many parks and protected areas in the Rocky Mountains are exploring solutions similar to Banff for resolving issues related to human, wildlife and vegetation interactions.

Lessons for collaboration and adaptive management

The Banff case history demonstrates the use of the recognized ingredients in successful adaptive ecosystem management and planning programmes (Clark, 1999; Cortner and Moote, 1999; Lal et al, 2001; Lee, 1993). A synergy created by a diverse and interested citizenry, informed with knowledge from a variety of sources, and interacting with scientists and managers, was the essential mechanism of an adaptive management programme that created innovative change (Hodgins et al, 2000). We conclude by discussing several important components of the BNP programme that may have application to restoration exercises in similar ecosystems.

Ecosystem restoration: Complexity, adaptive management and collaboration

The combined ecosystem-level impacts of modern human influences – such as fire suppression, highways and habitat fragmentation – are becoming increasingly recognized as very important in Rocky Mountain montane ecosystems (Baron, 2004; BBVS, 1996; Forman et al, 2003). The BNP experience showed that restoration effects on carnivores, herbivores and plants are likely to be complex, non-linear and dependent on starting conditions and neighbouring land effects. An initial analysis of response of indicators to restoration efforts to date (White et al, 2004) recommended that the maintenance and restoration of valley-bottom wildlife corridors and habitats should be the first priority in protected-area ecosystems affected by development (see Figure 12.6). If this landscape pattern is ecologically functional, fencing highways will probably result in an initial rise in ungulate populations due to reduced road-kill. This may be followed by an increasing density of more wary carnivores as road-caused mortality rates decline. At this stage, prey may utilize areas with high human use as refuges from predation, stabilizing wildlife populations at relatively high levels.

Where complete restoration occurs, and predator and prey sympatrically utilize habitats, predation may reduce ungulate abundance (Hebblewhite et al, 2005), thus reducing predator abundance. Prescribed burning to restore vegetation communities such as aspen and willow will be most successful when herbivores are limited to low densities by predation and, in long-term ecosystems, some human hunting (White et al, 1998; 2003).

The time lags and complexity in montane ecosystem change after restoration indicate that predicting and understanding the outcomes of costly mitigation action is still as much art as science. Because this is long-term, expensive work that often occurs where people live (near highways and towns), montane ecosystem restoration provides an excellent opportunity to involve stakeholders. Participatory activities may include monitoring (for example, wildlife use of crossing structures or wildlife corridors), joint scientist–stakeholder workshops, and advisory committees.

Integrating multidisciplinary knowledge with stakeholder collaboration

Sadly, true interdisciplinarity is uncommon between scientists, many of whom can barely communicate across disciplines due to isolation, jargon and perceptions of academic supremacy (Cortner and Moote, 1999). However, this integration can occur when the knowledge of multiple disciplines (for example, anthropology, history, traditional knowledge, ecology) is provided to a broad range of stakeholders (for example, park visitors, residents, First Nations and so on) with experience in civic discourse and collective learning (Clark, 1999; Shannon and Antypas, 1996).

When attempting to influence collaboration groups, scientists and other purveyors of specialized knowledge must learn to address useful problems, seek academic consensus across disciplines, develop logical predictions, communicate simply and avoid intellectual arrogance (Meffe and Viederman, 1995). In short, common sense and mutual respect are important. In the Banff case, such interactions between stakeholders, scientists and managers frequently led to deeper understanding of ecosystem change, and more innovative proposals for ecological restoration.

The long-term role of people

Interdisciplinary understanding increasingly suggests that small numbers of humans were probably an important long-term component of many ecosystems, through their activities as both hunters and burners. In contrast, today's millions of park visitors prefer to photograph wildlife against a backdrop of dense, green forests unobscured by smoke. This complete change in human-use levels and behaviour patterns is the underlying cause for many of today's ecosystem management challenges.

This was a revolutionary perspective in understanding Banff's ecosystem that, under the natural regulation paradigm, was perceived as a wilderness icon. Including people, both past and present, as active stewards of ecosystems provides a foundation for active adaptive management and participatory interest in ecological restoration programmes (Cortner and Moote, 1999). This change in perspective clearly requires engaging disciplines outside the pure natural sciences – such as anthropology, archaeology, traditional knowledge and local citizen perspectives.

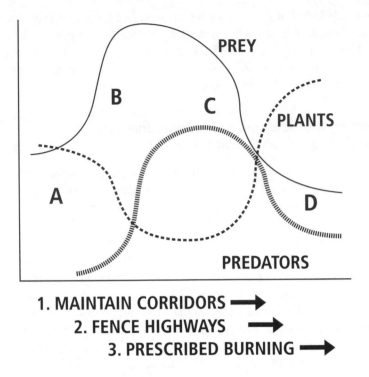

Figure 12.6 Stylized model for potential indicator response from highway wildlife mortality, habitat fragmentation and prescribed fire mitigation in montane eco-regions

Note: At A both ungulates and predator numbers are depressed by highway mortality. At B ungulates have responded to initial highway mitigation. At C, reduced highway mortality but ongoing high human use of wildlife corridors has created a prey refuge, which stabilizes ungulates and predators at relatively high numbers. At D, wildlife corridor restoration has removed the prey refuge.

Involvement of First Nations and use of traditional knowledge

If true interdisciplinary and multi-stakeholder approaches continue to guide ecological restoration, it will become increasingly important to continue to reach out to First Nations to bring their perspective and traditional knowledge into the consilience process (Parks Canada Agency, 2000). As one example, Banff National Park's prescribed burning is currently guided by modern technical standards for fire frequency and severity (White et al, 2005), but long-term burning by humans is an art that enhanced unique habitats for important wildlife and plant species (Pyne, 1995, 2004). As a second example, Banff is currently evaluating the feasibility of returning bison, once the dominant large mammal in east-slope montane ecosystems (Kay et al, 2000). This also will require not just the knowledge of modern science, but also the traditional knowledge of humans, one of bison's most important historic predators.

Conclusion

In the quarter century since 1980, the Bow Valley of Banff National Park and the adjacent province of Alberta has greatly changed. There are negative signs for the ecosystem: the resident human population has nearly tripled, park visitation has more than doubled, and the highway is twinned and busier than ever. But there are also many positive signs: wolves have returned and persisted, grizzly bear numbers are probably increasing and, largely due to the restored role of predators limiting ungulates, aspen and willow regeneration is widespread (Hebblewhite et al, 2005). Possibly most importantly, the level of citizen understanding and participation in ecosystem management issues is very high. Discourses over wildlife corridor width, the frequency of highway carnivore mortality, and the role of wildland fire in forest health take place almost daily in the newspapers, meeting rooms and coffee shops of the valley. The complexity of change over time clearly shows that ecological restoration of Canadian Rockies ecosystems will continue to be a journey of learning for stakeholders, researchers and managers alike. No discipline or interest group can claim primacy to lead this adventure into the future. All must contribute their knowledge and appreciation of this magnificent landscape to maintain it as a natural and cultural legacy for future generations.

Acknowledgements

Ideas for this paper were synthesized from four Banff case histories presented at the IRMMA conference. We appreciate the contributions of D. Dalman, H. Dempsey, S. Donelon, K. Endres, M. Hebblewhite, T. Hurd, C. Kay, D. Leighton, N. Maier, B. McFarlane, M. McIvor, D. McVetty, J. Otton, I. Pengelly, D. Peterson, B. Ritchie, J. Roulet, R. Sandford and P. Vienotte to the case histories, and of M. Price and L. Taylor for conference organization.

References

Alberta Economic Development and Tourism (1994) *Economic Income of Visitors to Banff National Park in 1991*, Edmonton, Alberta Economic Development and Tourism
Baker, W. L., Munroe, J. A. and Hessl, A.E. (1997) 'The effects of elk on aspen in the winter range in Rocky Mountain National Park', *Ecography*, vol 20, pp155–165
Balsiger, P. W. (2004) 'Supradisciplinary research practices: History, objectives, and rationale', *Futures*, vol 36, pp407–421
Banff Heritage Tourism Corporation (2004) *Strategic Plan 2004–2005*, Banff, Banff Heritage Tourism Corporation
Baron, J. S. (2004) *Rocky Mountain Futures: An Ecological Perspective*, Washington DC, Island Press
BBVS (Banff Bow Valley Study) (1996) *Banff-Bow Valley: At the Crossroads*, Technical Report of the Banff-Bow Valley Task Force, Ottawa, Department of Canadian Heritage
Bella, L. (1987) *Parks for Profit*, Montreal, Harvest House

Benn, B. and Herrero, S. (2002) 'Grizzly bear mortality and human access in Banff and Yoho National Parks 1971–1998', *Ursus*, vol 13, pp213–221

Bow Corridor Ecosystem Advisory Group (1998) *Wildlife Corridor and Habitat Patch Guidelines for the Bow Valley*, Municipal District of Bighorn, Town of Canmore, Banff National Park and the Alberta Government

Bow Corridor Ecosystem Advisory Group (1999) *Guidelines For Human Use Within Wildlife Corridors and Habitat Patches in the Bow Valley (Banff National Park to Seebee)*, Municipal District of Bighorn, Town of Canmore, Banff National Park and the Alberta Government

Caughley, G. (1979) 'What is this thing called carrying capacity?', in Boyce, M. S. and Hayden-Wing L. D (eds) *North American Elk: Ecology, Behaviour, and Management*, Laramie, University of Wyoming Press

Chruszcz, B., Clevenger, A. P., Gunson, K. E. and Gibeau M. L. (2003) 'Relationships among grizzly bears, highways, and habitat in the Banff-Bow Valley, Alberta, Canada', *Canadian Journal of Zoology*, vol 81, pp1378–1391

Clark, T. W. (1999) 'Interdisciplinary problem-solving: Next steps in the greater Yellowstone ecosystem', *Policy Sciences*, vol 32, pp393–414

Clevenger, A. P. and Waltho, N. (2000) 'Factors influencing the effectiveness of wildlife underpasses in Banff National Park, Alberta, Canada', *Conservation Biology*, vol 14, pp47–56

Clevenger, A. P., Chruszcz, B. and Gunson, K. (2001) 'Highway mitigation fencing reduces wildlife-vehicle collisions', *Wildlife Society Bulletin*, vol 29, pp646–653

Clevenger, A. P., Chruszcz, B., Gunson, K. and Wierzchowski, J. (2002) *Roads and Wildlife in the Canadian Rocky Mountains: Movements, Mortality and Mitigation*, Final Report to Parks Canada, Banff, Parks Canada

Cole, G. F. (1971) 'An ecological rationale for the natural or artificial regulation of native ungulates in parks', *Transactions of the North American Wildlife Conference*, vol 36, pp417-425

Cooper, B., Hayes, J. and LeRoy, S. (2002) *Science Fiction or Science Fact? The Grizzly Biology behind Parks Canada Management Models*, Vancouver, BC, Fraser Institute

Cornwell, L. and Costanza, R. (1996) 'A futures outlook of the Banff Bow Valley: A modelling approach to ecological and social issues', in Green, J., Pacas, C., Bayley S. and Cornwell, L. (eds) *A Cumulative Effects Assessment and Futures Outlook for the Banff-Bow Valley*, Ottawa, Department of Canadian Heritage

Cortner, H. J. and Moote, M. A. (1999) *The Politics of Ecosystem Management*, Washington DC, Island Press

Cowan, I. M. (1947) 'The timber wolf in the Rocky Mountain parks of Canada', *Canadian Journal of Research*, vol 25, pp139–174

Draper, D. (2000) 'Toward sustainable mountain communities: Balancing tourism development and environmental protection in Banff and Banff National Park, Canada', *Ambio*, vol 29, pp408–415

Duke, D., Hebblewhite, M., Paquet, P., Callaghan, C. and Percy, M. (2001) 'Restoration of a large carnivore corridor in Banff National Park', in Maehr, D., Noss, R. F. and Lavlin, J. L. (eds) *Large Mammal Restoration: Ecological and Social Challenges in the 21st Century*, Washington DC, Island Press

Eagles, P. F. J. (2002) 'Environmental management', in Dearden, P. and. Rollins, R. (eds) *Parks and Protected Areas in Canada*, Oxford, Oxford University Press

Endter-Wada, J. D., Blahna, D., Krannich, R. and Brunson, M. (1998) 'A framework for understanding social science contributions to ecosystem management', *Ecological Applications*, vol 8, pp891–904

Fedje, D. W., White, J. W., Wilson, M. C., Nelson, D. E., Vogel, J. S. and Southon, J. R. (1995) 'Vermilion Lakes site: Adaptions and environments in the Canadian Rockies during the late Pleistocene and early Holocene', *American Antiquity*, vol 60, pp81–108

Forman, R. T. T., Bissonnette, J., Clevenger, A., Cutshall, C., Dale, V., Fahrig, L., Goldman, C., Heanue, K., Jones, J., Sperling, D., Swanson, F., Turrentine, T. and Winter, T. (2003) *Road Ecology: Science and Solutions*, Washington DC, Island Press

Gibeau, M. L., Herrero, S., Kansas, J. and Benn, B. (1996) 'Grizzly bear population and habitat status in Banff National Park', in Green, J., Pacas, C., Bayley, S. and Cornwell L. (eds) *A Cumulative Effects Assessment and Futures Outlook for the Banff-Bow Valley*, Ottawa, Department of Canadian Heritage

Gibeau, M. L., Herrero, S., McLellan, B. N. and Woods, J. G. (2001) 'Managing for grizzly bear security areas in Banff National Park and the central Canadian Rocky Mountains', *Ursus*, vol 12, pp121–130

Gibeau, M. L., Clevenger, A. P., Herrero, S. and Wierzchowski, J. (2002) 'Grizzly bear response to human development and activities in the Bow River watershed, Alberta, Canada', *Biological Conservation*, vol 103, pp227–236

Golder and Associates (1994) *Environmental Screening: Middle Springs II Area Structure Plan*, Banff, Town of Banff

Green, J., Pacas, C., Bayley, S. and Cornwell, L. (eds) (1996) *A Cumulative Effects Assessment and Futures Outlook for the Banff-Bow Valley*, Ottawa, Department of Canadian Heritage

Hamer, D. (1996) 'Buffaloberry (*Shepherdia canadensis* (L.) Nutt.) fruit production in fire successional bear feeding sites', *Journal of Range Management*, vol 49, pp520–529

Hamer, D. (1999) 'Forest fires influence on yellow hedysarum habitat and its use by grizzly bears in Banff National Park, Alberta', *Canadian Journal of Zoology*, vol 77, pp1513–1520

Hart, E. J. (2003) *The Battle for Banff: Exploring the Heritage of the Banff-Bow Valley, Part 2. 1930 to 1985*, Altona, Manitoba, Friesens

Hawkes, B. C. (1980) 'Fire history in Kananaskis Provincial Park, Alberta', in *Proceedings of the Fire History Workshop*, USDA Forest Service General Technical Report RM-81, Fort Collins, Rocky Mountain Forest Research Station

Hebblewhite, M., Pletscher, D. H. and Paquet, P. C. (2002) 'Elk population dynamics in areas with and without wolves in Banff National Park', *Canadian Journal of Zoology*, vol 80, pp789–799

Hebblewhite, M., White, C. A., Nietvelt, C., McKenzie, J. M., Hurd, T. E., Fryxell, J. M., Bayley, S. and Paquet, P. C. (2005) 'Human activity mediates a trophic cascade caused by wolves', *Ecology*, vol 86, pp2135–2144

Herrero, S., Roulet, J. and Gibeau, M. L. (2001) 'Banff National Park: Science and policy in grizzly bear management', *Ursus*, vol 12, pp161–168

Hess, K. (1993) *Rocky Times in Rocky Mountain National Park*, Niwot, University Press of Colorado

Hessl, A. (2002) 'Aspen, elk, and fire: The effects of human institutions on ecosystem processes', *BioScience*, vol 52, pp1011–1022

Hodgins, D. W., Green, J. E., Harrison, G. and Roulet, J. (2000) 'From confrontation to conservation: The Banff National Park experience', in *Wilderness Science in a Time of Change Conference*, USDA Forest Service Proceedings RMRS-P-15, vol 2, Ogden, Rocky Mountain Forest Research Station

Holroyd, G. S. and Van Tighem, K. J. (1983) *Ecological (Biophysical) Land Classification of Banff and Jasper National Parks. Volume III: Wildlife Inventory*, Edmonton, Canadian Wildlife Service

Huff, D. E. and Varley, J. D. (1999) 'Natural regulation in Yellowstone National Park's northern range', *Ecological Applications*, vol 9, pp17–29

Jamal, T. B. and Getz, D. (1999) 'Community roundtables for tourism related conflicts', *Journal of Sustainable Tourism*, vol 7, pp356–378

Jamal, T. B., Steink, S. M. and Harper, T. L. (2002) 'Beyond labels: Pragmatic planning in multistakeholder tourism-environmental conflicts', *Journal of Planning*

Education and Research, vol 22, pp164–177

Johnson, E. A., Miyanshi, K. and Weir, J. M. H. (1995) 'Old-growth, disturbance, and ecosystem management', *Canadian Journal of Botany*, vol 73, pp918–926

Johnson, E. A. and Larsen, C. P. S. (1991) 'Climatically induced change in fire frequency in the southern Canadian Rockies', *Ecology*, vol 72, pp194–201

Jones, K. R. (2002) *Wolf Mountains: A History of Wolves along the Great Divide*, Calgary, University of Calgary Press

Kay, C. E. (1998) 'Are ecosystems structured from the top-down or the bottom-up: A new look at an old debate', *Wildlife Society Bulletin*, vol 26, pp484–498

Kay, C. E. and White, C. A. (1995) 'Long-term ecosystem states and processes in the central Canadian Rockies: A new perspective on ecological integrity and ecosystem management', *George Wright Society Annual Conference on Research and Resources in Parks and on Protected Lands*, vol 8, pp119-132

Kay, C. E., White, C. A, Pengelly, I. R. and Patton B. (1999) *Long-term Ecosystem States and Processes in Banff National Park and the Central Canadian Rockies*, Parks Canada Occasional Report 9, Ottawa, National Parks Branch

Kay, C. E., Patton B. and White, C. A. (2000) 'Historical wildlife observations in the Canadian Rockies: Implications for ecological integrity', *Canadian Field Naturalist*, vol 114, pp561–583

Kloppers, E. L., Cassidy St. Clair, C. and Hurd, T. E. (2005) 'Predator-resembling aversive conditioning for managing habituated wildlife', *Ecology and Society*, vol 10, no 1, p31, www.ecologyandsociety.org/vol10/iss1/art31/

Komex International (1995) *Atlas of the Central Rockies Ecosystem*, Calgary, Central Rockies Ecosystem Interagency Liaison Group

Lal, P., Lim-Applegate, H. and Scoccimaro, M. (2001) 'The adaptive decision-making process as a tool for integrating natural resource management', *Conservation Ecology*, vol 5, no 2, p11, www.consecol.org/vol5/iss2/art11/

Landres, P. B., Morgan, P. and Swanson, F. J. (1999) 'Overview of the use of natural variability concepts in managing ecological systems', *Ecological Applications*, vol 9, pp1179–1188

Lee, K. N. (1993) *Compass and Gyroscope*, Washington DC, Island Press

McFarlane, B. L., Stumpf-Allen, R. C. G. and Watson, D. O. (2004) *Managing for Mountain Pine Beetle in Kootenay and Banff National Parks: A Survey of Park Visitors and Local Residents*, Edmonton, Canadian Forest Service, Northern Forestry Centre

McGuire, T. M. and Morrall, J. F. (2000) 'Strategic improvements to minimize highway environmental impacts within the Canadian Rockies national parks', *Canadian Journal of Civil Engineering*, vol 27, pp523–532

McKenzie, J. A. (2001) *The Selective Advantage of Urban Habitat Use by Elk in Banff National Park*, MSc thesis, Guelph, University of Guelph,

Meffe, G. M. and Viederman, S. (1995) 'Combining science and policy in conservation biology', *Wildlife Society Bulletin*, vol 23, pp327–332

Morgan, P., Aplet, G. H., Haufler, J. B., Humphries, H. C., Moore, M. M. and Wilson, W. D. (1994) 'Historical range of variability: A useful tool for evaluating ecosystem change', *Journal of Sustainable Forestry*, vol 2, pp87–111

Mountain Parks Visitor Survey Partnership (2004) *2003 Survey of Visitors to Banff, Jasper, Kootenay, and Yoho Parks of Canada: Summary Report*, Winnipeg, Parks Canada Western Service Centre

Nelson, J. G. and Byrne, A. R. (1966) 'Man as an instrument of landscape change: Fires, floods, and national parks in the Bow Valley, Alberta', *Geographical Review*, vol LVI, no 2, pp226–238

Nietvelt, C. G. (2001) *Herbivory Interactions between Beaver* (Castor canadensis*) and Elk* (Cervus elaphus*) on Willow* (Salix *spp.) in Banff National Park, Alberta*, MSc Thesis, Edmonton, University of Alberta

Pacas, C., Bernard, D., Marshall, N. and Green, J. (1996) *State of the Banff Bow Valley*, Prepared for the Banff Bow Valley Study, Ottawa, Department of Canadian Heritage

Paquet, P., Wierzchowski, J. and Callaghan, C. (1996) 'Effects of human activity on gray wolves in the Bow River Valley, Banff National Park, Alberta', in Green, J., Pacas, C., Bayley, S. and Cornwell, L. (eds) *A Cumulative Effects Assessment and Futures Outlook for the Banff-Bow Valley*, Ottawa, Department of Canadian Heritage

Parks Canada (1997) *Banff National Park Management Plan*, Ottawa, Department of Canadian Heritage

Parks Canada (1999) *Environmental Screening: Elk Management Strategy in the Bow Valley, Banff National Park*, Banff, Banff National Park

Parks Canada (2002) *Environmental Screening: Regional Forest Management Strategy in Banff National Park and Adjacent Alberta Lands*, Banff, Banff National Park

Parks Canada Agency (2000) *Unimpaired for Future Generations? Conserving Ecological Integrity with Canada's National Parks*, Report of the Panel on the Ecological Integrity of Canada's National Parks, Ottawa, Department of Canadian Heritage

Petersen, D. (2000) 'Grizzly bears as a filter for human use management in Canadian Rocky Mountain national parks', in *Wilderness Science in a Time of Change Conference*, USDA Forest Service Proceedings RMRS-P-15, vol 5, Ogden, Rocky Mountain Forest Research Station

Pyne, S. J. (1995) *World Fire: The Culture of Fire on Earth*, New York, NY, Henry Holt and Company

Pyne, S. J. (2004) 'Burning Banff', *Interdisciplinary Studies in Literature and Environment*, vol 11, pp221–247

Rhemtulla, J. M., Hall, R. J., Higgs, E. S. and Macdonald, S. E. (2002) 'Eighty years of change: Vegetation in the montane ecoregion of Jasper National Park, Alberta, Canada', *Canadian Journal of Forest Research*, vol 32, pp2010–2021

Ripple, W. J., Larsen, E. J., Rankin, R. A. and Smith, D.W. (2001) 'Trophic cascades among wolves, elk, and aspen on Yellowstone's northern range', *Biological Conservation*, vol 102, pp227–234

Ritchie, J. B. R. (1999) 'Crafting a value-driven vision for a national tourism treasure', *Tourism Management*, vol 20, pp273–282

Shannon, M. A. and Antypas, A. R. (1996) 'Civic science is democracy in action', *Northwest Science*, vol 70, pp66–99

Tande, G. F. (1979) 'Fire history and vegetation pattern of coniferous forests in Jasper National Park, Alberta', *Canadian Journal of Botany*, vol 57, pp1912–1939

White, C. A. (1985) *Wildland Fires in Banff National Park 1880–1980*, Occasional Report 3, Ottawa, Parks Canada

White, C. A., Olmsted, C. E. and Kay, C. E. (1998) 'Aspen, elk, and fire in the Rocky Mountain national parks of North America', *Wildlife Society Bulletin*, vol 26, pp449-462

White, C. A., Feller, M.–C. and Vera, P. (2001) 'New approaches for testing fire history hypotheses', *Proceedings of the International Conference on Science and Management of Protected Areas*, vol 4, pp398–411

White, C. A., Feller, M. C. and Bayley, S. (2003) 'Predation risk and the functional response of elk-aspen herbivory', *Forest Ecology and Management*, vol 181, pp77–97

White, C. A., Hurd, T. E., Hebblewhite, M. and Pengelly, I. R. (2004) 'Mitigating fire suppression, highway, and habitat fragmentation effects in the Bow Valley ecosystem, Banff National Park', Paper presented at *Monitoring the Effectiveness of Biological Conservation Conference*, Vancouver, FORREX and University of British Columbia

White, C. A., Pengelly, I. R., Zell, D. and Rogeau, M. P. (2005) 'Restoring heterogeneous fire regimes in Banff National Park', in Taylor, L., Zelnik, J., Cadwallader, S.
and Hughes B. (eds) *Mixed Severity Fire Regimes: Ecology and Management*,
Association of Fire Ecology, Pullman, Washington State University Extension,
pp255–266

Wierzchowski, J., Heathcott, M. and Flannigan, M. D. (2002) 'Lightning and lightning
fire, central cordillera, Canada', *International Journal of Wildland Fire*, vol 11,
pp41–51

Wight, P. (2001) *Integration of Biodiversity and Tourism: Canada Case Study*, Report to
UNEP's Biodiveristy Planning Support Programme, www.unep.org/bpsp/
Tourism/Case%20 Studies%20(pdf)/CANADA%20(Tourism).pdf

Woods, J. G. (1990) *Effectiveness of Fences and Underpasses on the Trans-Canada
Highway and their Impact on Ungulate Populations*, Calgary, Environment Canada,
Canadian Parks Service, Western Regional Office

Woods, J. G., Cornwell, L., Hurd T., Kunelius, R., Paquet, P. and Wierzchowski, J.
(1996) 'Elk and other ungulates', in Green, J., Pacas, C., Bayley, S. and Cornwell,
L. (eds) *A Cumulative Effects Assessment and Futures Outlook for the Banff-Bow
Valley*, Ottawa, Department of Canadian Heritage

Zinkan, C. and Syme, I. (1997) 'Changing dimensions of park management', *Forum for
Applied Research and Public Policy*, vol 12, pp39–42

13

Integrated Restoration and Rehabilitation of Powerline Corridors in Mountain National Parks in Australia

Stuart Johnston and Roger Good

Introduction

This chapter considers an area within three extensive and contiguous national parks, Kosciuszko National Park and Brindabella National Park in New South Wales (NSW) and Namadgi National Park in the Australian Capital Territory (ACT), as well as two large commercial state forests. The parks and forest lands are part of the landscape of the Snowy Mountains and Brindabella Ranges (part of the Australian Alps), the major water catchments in southeastern Australia. The project area covers six high-voltage powerline easements that traverse the mountains as an extensive corridor over a distance of some 300km. The easements are the responsibility of TransGrid, a state-owned corporation that operates and manages the NSW high voltage electricity transmission system. One of TransGrid's primary activities is to maintain transmission-line easements to ensure the safe and reliable supply of electricity for the state of NSW. In particular, these activities involve the ongoing management of vegetation to ensure suitable clearance from the powerlines. As the easement corridors of this project occur within and dissect national parks and forests, they are a major management consideration for both the power transmission agency (TransGrid) and the national parks and forestry land management agencies (see Figure 13.1).

The project

In early 2001, as part of vegetation maintenance programmes, contractors cleared extensive areas of native vegetation along two transmission-line easements traversing the three national parks. As a response, a project was initiated in June 2001 to rehabilitate excessively damaged areas and to identify sustainable ecological management practices that could be used in the Australian Alps and transferred to other areas across NSW. To protect the integrity of the unique mountain environment and ecosystems for all interest groups, it is vital to have an integrated management framework that identifies both the use of these areas for electricity generation and transmission, and the habitat types and conservation values of ecosystems, and involves stakeholders in decision-making.

Aims and objectives of the project

The aim of the project is to ensure that the extensively disturbed areas along the easement corridors are rehabilitated to a near-natural and functional ecosystem condition that can be maintained in the future, predominantly through the enhancement of ecological processes. This aim has been, and will be, achieved through the following objectives:

- to establish an initial vegetation cover, which provides primary soil stability and the progressive establishment (over-planting) and natural recruitment of a native herbaceous and shrub complex that renders the easements ecologically stable;
- to establish a maintenance programme based on the enhancement of ecological processes, which:
    - removes the need for use of heavy equipment in maintenance programmes;
    - enhances exotic species management along the easement as part of ecological maintenance and management;
    - identifies and enables implementation of other easement maintenance techniques that do not impact unacceptably on the natural environment;
    - promotes the establishment of a vegetation structure that provides for native fauna movement across the easements;
- to develop and implement TransGrid maintenance plans and programmes in conjunction with all stakeholders to ensure mutually acceptable outcomes that comply with all environmental planning, conservation and electricity supply legislation;
- to use the information gathered from this project to target community conservation activities in NSW (outside national parks), focusing on the loss of biodiversity, deteriorating water quality, soil acidity and increasing salinity in NSW catchments by:
    - working with farmers and landholders in practical on-the-ground management activities across the landscape;
    - vegetation rehabilitation through joint Greening Australia and Landcare Australia ecological restoration activities.

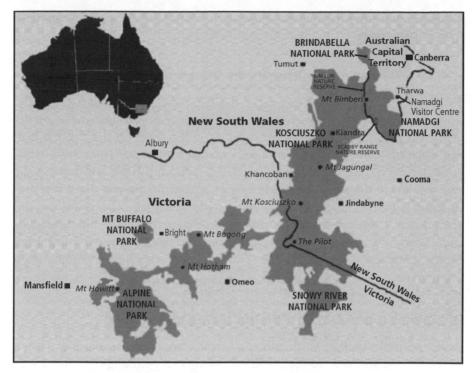

Figure 13.1 Location of Brindabella and Kosciuszko National Parks
in relation to the Australian Alps

Time-frame and funding sources

The project commenced in March 2001 with the planning and implementation
of the initial rehabilitation work. As new techniques were developed in relation
to assessment, rehabilitation and monitoring over the subsequent three years,
the project moved into a second stage, that of integrating the knowledge
gained into other environments and rehabilitation activities across the state.

The dissemination of the information gained from this project is continuing through lectures, workshops and field days with industry, government
agencies and community groups in conjunction with organizations such as
Greening Australia, Landcare Australia and the Australian Network for Plant
Conservation (ANPC), with ongoing funding committed for these activities
until at least 2009.

The funding for the entire programme (approximately AU$7 million, or
US$6 million) has been provided by TransGrid with other agencies and orga-
nizations – NSW National Parks and Wildlife Service (NPWS), Department of
Infrastructure, Planning and Natural Resources (DIPNR), ANPC, Greening
Australia, the Australian National University (ANU), NSW State Forests

(NSWSF) and the Environment Protection Agency (EPA) – providing long-term in-kind, research and management staff support.

Interdisciplinarity in the project team

To define the aims and objectives of the rehabilitation programme and the subsequent ecological maintenance and community programmes to be implemented as part of the project, a consultative project team (steering committee) was established. It consisted of stakeholder experts from TransGrid, NSW NPWS, NSWSF, Greening Australia, ANPC, ANU and NSW Soil Services. Its task was to consider ongoing performance against the project objectives and issues of concern relating to planned future activities. The positions and composition of the committee continually evolved as the project changed focus and moved from a works-based project to one of dissemination and integration of the knowledge and expertise gained during the initial rehabilitation programme.

Disciplines represented in the project team

As the project required information and inputs from many ecological, natural area land management, forestry and environmental planning disciplines – as well as specific data on soils, vegetation, climatic regimes, biodiversity and rehabilitation techniques and materials – a wide diversity of interests and skills were and are represented in the multidisciplinary project team. These are:

- soil condition and slope stability – ANU, TransGrid and NSW Soil Services;
- native vegetation – NSW NPWS, Environment ACT and Greening Australia;
- water quality and stream health – Cooperative Research Centre for Freshwater Ecology;
- fauna – NSW NPWS Threatened Species Unit;
- fish – Murray Darling Basin Commission, Cooperative Research Centre for Freshwater Ecology and Environment ACT;
- cultural heritage – archaeological consultants and Aboriginal Land Councils, NSW NPWS;
- rehabilitation – NSW Soil Services, NSW NPWS, TransGrid and ANU;
- electricity industry – TransGrid;
- environmental planning – NSW NPWS, NSW EPA, TransGrid, ANPC and Environment ACT.

Mechanisms to foster interdisciplinary concepts

Early in the establishment of the project, the project team identified many issues that mitigated against cooperative and committed input by a diverse group of experts, and which would therefore require effective direction, organization and management to successfully meet the identified aims and objectives. To draw together the diversity of experts and to integrate their individual skills and knowledge, the project team pursued a number of initiatives to foster and ensure full collaboration and input from all disciplines. These initiatives included:

- the preparation of an initial project scoping plan and a detailed site-by-site rehabilitation plan;
- the appointment of a project manager;
- interdisciplinary stakeholder field inspections and planning days;
- structured and planned collaboration between the relevant agencies and research organizations in the implementation of the works programmes;
- a planned and engendered ownership of the project by individual stakeholders and by all contributing external community agencies and groups;
- a planned and structured collaborative input to, and preparation of, all plans, training programmes, workshops, reports, manuals and technical and scientific publications;
- the establishment of an electronic 'rehabilitation network' for all interested parties to discuss and contribute to the project on an informal but regular basis;
- interdisciplinary workshops in collaboration with ANPC and other stakeholder groups;
- extensive use of a geographic information system (GIS) providing a capacity to integrate, manipulate and map information and data provided by stakeholders and other contributing organizations and individual specialists;
- establishment of a long-term monitoring programme to continually assess the success of the soil stability, vegetative cover and ecological function. The outcomes of the monitoring are a direct input to post-rehabilitation maintenance and programmes that evolved from this research.

The challenges, successes and failures of the interdisciplinary project team

Challenges

As with many major projects with a diversity of stakeholder contributions, the greatest challenge for the project team has been to engender and maintain a high level of effective and continuous communication between all stakeholders. This is necessary to avoid contrasts and conflicts in:

- individual stakeholders' perceptions of the issues and problems associated with the project;
- individual stakeholders' perceptions of the desired outcomes of the project;

- individual stakeholders' priorities in terms of time-frames and activities, and to ensure:
    - maintenance of equity in the implementation of the project;
    - maintenance of a full and continuing commitment of the stakeholders to the project over its full term (until completion in 2008).

Successes to date

The successes of the interdisciplinary team approach to the rehabilitation project can be recognized in terms of the stability of the rehabilitated area, the biodiversity that has been restored and developed, and the continuing enthusiasm exhibited by all stakeholders to see the rehabilitation work through to one of a stable self-sustaining ecological community, albeit a recreated landscape and ecosystem.

The community involvement and subsequent community-based environmental programmes and partnerships, and the integration of the findings of the project into new management strategies and programmes across the state can also be identified as major successes and outcomes of the project.

Failures

There were no notable failures but, as with all rehabilitation projects, several small site-specific failures to achieved a stable vegetation complex occurred. Through the continuous monitoring and assessment process, these were identified as resulting from a limited understanding of the actual degree and extent of the degradation of the natural environments (soil structure, nutrients and soil fauna) that the project team had to address. Fortunately these were only small area failures and were effectively addressed. They also served to further the team's understanding of ecological functionality and stability, and the ecological rehabilitation techniques necessary to ensure their full restoration.

In terms of the management of the project, matching community and individual stakeholder demands and issues along with legislative requirements was at times challenging for the project team.

Stakeholder involvement

Definition of issues, priorities and activities

As outlined above, the project team included representatives from TransGrid and the government agencies whose legislative responsibilities and obligations covered the easement clearing and subsequent rehabilitation works. All initial policy, planning, works planning and priorities were identified and set by the project team.

The project team also identified other areas of expertise that were likely to be drawn upon for the project to progress, and established formal and informal protocols with appropriate and relevant agencies, community groups and personnel to ensure these were part of the planning and ongoing implementation of the programme. Examples include experts in hydrology and catchment management, and ornithologists.

The issues, priorities and activities were defined under legislative obligations and constraints but were broadened by issues such as research requirements, knowledge base and exchange, ecological understanding, skills and techniques, monitoring requirements (legal and ecological), the day-to-day field implementation programme, and teaching and information dissemination techniques.

Implementation of research

An initial task of the project team was to identify existing or potential gaps in the understanding of ecological rehabilitation knowledge and techniques (Good, 2000, 2004, 2006) relevant to the project and to seek appropriate research personnel to address or contribute to this understanding. A number of specific and relevant research projects were carried out prior to and/or in conjunction with the field rehabilitation programme (Good and Johnston, 2001) by university and state government agency research personnel and the Commonwealth Scientific and Industrial Research Organisation (CSIRO). These covered aspects of soil nutrient availability (Good and Johnston, 2004; Johnston, 1998; Johnston and Good, 1996; Johnston and Ryan, 2000) and plant nutrition (Good and Johnston, 2004), mulching and fertilizer rates, native plant species propagation, landscape function analysis, ecological state and transition modelling (Johnston et al, 2003), fauna habitat assessment and biodiversity survey (Johnston and Johnston, 2003, 2004). Many of the research projects are ongoing and will contribute to other major ecological landscape rehabilitation and revegetation programmes in the future. The research projects identified by the project team as essential to the programme were, and continue to be, reviewed on a regular basis through the project team's planning and assessment meetings.

Results and outcomes

The extensive clearing and subsequent rehabilitation programme have provided an ideal opportunity to develop and assess a number of new ecologically based rehabilitation techniques and to develop ecologically based maintenance programmes for the rehabilitated easement areas, which will avoid any further use of heavy machinery in the fragile and sensitive mountain soil and vegetation systems (Johnston and Good, 2004). The project has also contributed to an understanding of how physical, chemical and biological components of the environment are linked, with a focus on impacts involving the electricity industry, protected area management, rare and endangered species, ecosystem function, and commercial and recreational uses.

The monitoring programme (see Figure 13.2) has also provided an insight into which native plant species grow and survive best under environmentally stressed conditions and therefore will be the major species used for future rehabilitation work in similar high mountain environments. The sharp vegetation and habitat boundaries that now exist between the natural bushland and the cleared easements also provide an opportunity to study the dynamics of the ecotone between the stable natural bushland, which borders the powerline easements, and the cleared and rehabilitated areas along the easements. The long-term monitoring of the rehabilitation and revegetation work has enabled, and will enable, ecologically based management programmes to be developed and implemented as standard powerline easement management in other localities across which the powerline grid extends. These include:

- a model to assess the cumulative impacts of multiple use and the identification of degradation thresholds;
- methods to assess the effects of human uses on conservation and other values of the ecosystems of the Australian Alps;
- strategies based on adaptive management in response to the results of monitoring programmes, including new procedures and processes for transmission-line easement management and maintenance, with emphasis on protected or sensitive lands.

Much has been gained from this opportunity to apply ecological rehabilitation and restoration principles to the revegetation programme and to develop complementary management/maintenance programmes and guidelines for both the high-voltage transmission-line easements across the state and several threatened native plant and animal species. Habitat for specific species has been reconstructed, as well as a number of different structural plant species corridors across the easements for use by native animals and birds. These will be managed as an integral part of the easement and national park management.

The project has also recognized the need for government agencies and research institutes to collaborate. The results of the research and survey programmes have been used by the individual agencies and organizations represented in the interdisciplinary project team in implementing their own revegetation and land management programmes. A rehabilitation network has evolved and has widely disseminated the results of the research through the project team to other local, national and international research and land management organizations. At a local and national level, the rehabilitation programme and the results of associated programmes have been disseminated by reports, media releases, field days, rehabilitation workshops and hands-on training programmes, scientific papers and through university lectures on landscape rehabilitation and management. A widely acknowledged example is that of the 'Ecological Restoration Workshops' presented for practitioners, government agency staff and community groups that have been instigated as a result

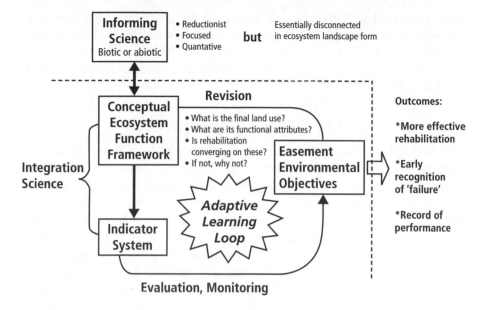

Figure 13.2 Monitoring procedure utilizing state and transition theory, and landscape function analysis

of this project and from which an ecological rehabilitation manual has been produced (Good, 2006).

The project has also led to a direct relationship between TransGrid and Greening Australia in the southern region of NSW through 'GreenGrid', a joint initiative/partnership between the two organizations, with the aim of 'Protecting and connecting habitat for threatened woodland birds, while simultaneously managing native vegetation to achieve multiple natural resource management objectives'. This partnership has enabled a new, targeted community conservation involvement in southern NSW, which now focuses on bridging the gap between existing Greening Australia sites and TransGrid's developing transmission network. The GreenGrid project is focused on several major river catchments in southern NSW, where loss and fragmentation of native vegetation have had negative impacts on biodiversity, and where water quality and soil salinity issues have been identified by research agencies and catchment management authorities as urgent priorities for attention. The majority of this work has been facilitated by working with farmers in practical, on-the-ground actions across the landscape.

A further significant outcome of the rehabilitation programme has been the forging of a close association between TransGrid, Greening Australia and the ANPC to develop and deliver environment-specific ecological rehabilitation training packages and workshops (Brown et al, 2003). Several have already been conducted with participants from industry, government agencies, non-government conservation groups, educational institutions, environmental consultants and interested individuals. The NSW State Government Environment Trust has provided additional supporting funds for further workshops in 2005 to 2007.

Implications of the project

Within TransGrid, the project has had a pronounced effect on all of its activities undertaken on transmission-line easements (especially easement maintenance activities) across the state. The project has changed attitudes within a predominantly structural engineering organization, from little concern for environmental issues to one where all planning, construction and maintenance activities are developed and implemented according to sound environmental planning and ecologically sustainable principles. The project has also resulted in many new standards, procedures, guidelines and agreements developed and written for TransGrid and the electrical industry in general, which assist in determining the relative scale of the environmental risks of its activities, products and services, so that these can be prioritized and managed through each company's environmental management system (see Box 13.1).

Box 13.1 Standards, guidelines, policies and agreements developed as a result of the project

TransGrid guidelines, policies and agreements

- 'Easement Planting with Native Birds in Mind' (see Figure 13.3)
- Environmental Assessment of Maintenance Works on Transmission Lines, Easements and Access Tracks
- TransGrid Environmental Manual
- TransGrid Guide to Environmental Legislation
- Inspection and Maintenance of Transmission Lines, Easements and Access Tracks
- A memorandum of understanding between the New South Wales National Parks and Wildlife Service and TransGrid for the inspection and maintenance of tracks and TransGrid infrastructure in national park
- A memorandum of understanding between the Australian Capital Territory and TransGrid for the inspection and maintenance of tracks and TransGrid infrastructure in national parks

Electricity industry in general

- Guidelines for the Control of Tree Growth
- NSW Environmental Law Manual for Electrical Industries
- Electricity Transmission and Distribution Industries, Guidelines for Easement Maintenance

Greening Australia

- Brochure on Bringing Native Birds Back (Greening Australia, 2004)

Figure 13.3 'Easement Planting with Native Birds in Mind'

Note: 'Easement Planting with Native Birds in Mind' is a series of brochures that have been developed in conjunction with Greening Australia to assist landholders when selecting plants for safe and environmentally suitable revegetation of transmission-line easements

Lessons for integrated research and management

From this rehabilitation project, many lessons in integrated interdisciplinary research and interagency collaboration have been learned and a set of guidelines/steps for the implementation of ecologically based rehabilitation programmes for both industry and conservation agencies have been developed. These include:

- facilitate dialogue between the relevant agencies/organizations;
- appoint a steering committee (its composition should be allowed to change over the duration of the project depending upon evaluation/needs);
- identify and clarify the actual issues/problems;
- identify and define the desired outcomes of the project;
- develop a project plan/guidelines to identify how to meet the desired outcomes;
- implement the plan and programme applying the 'precautionary principle' and recognizing the role of 'adaptive management';
- during the implementation of the plan, ensure that there is constant evaluation to identify if the desired outcomes are being met or are still relevant;
- ensure continual stakeholder involvement/inclusion throughout the implementation of the project;
- disseminate lessons learned from the project to a relevant broader audience.

References

Brown, C. L., Hall, F. and Mill, J. (eds) (2003) *Plant Conservation: Approaches and Techniques from an Australian Perspective*, Canberra, Australian Network for Plant Conservation

Good, R. B. (2000) 'Rehabilitation and revegetation of the Kosciuszko Summit Area: An historic review', in Mill, J. (ed.) *Proceedings of the 3rd Australian Network for Plant Conservation Conference, Albury*, Canberra, Australian Network for Plant Conservation

Good, R. B. (2004) 'Rehabilitating fire-damaged wetlands in the Snowy Mountains', *Special Edition – Rehabilitation, Australasian Plant Conservation*, vol 12, no 4, pp3–5

Good, R. B. (2006) *The Australian Alps Rehabilitation Manual: A Guide to Ecological Rehabilitation in the Australian Alps*, Canberra, Australian Alps Liaison Committee

Good, R. B. and Johnston, S. W. (2001) 'Ecological restoration in the Australian Alps: Two case studies – TransGrid and Snowy Mountains case studies', in Mill, J. (ed.) *Proceedings of the 3rd Australian Network for Plant Conservation Conference, Albury*, Canberra, Australian Network for Plant Conservation

Good, R. B. and Johnston, S. W. (2004) 'Ecological restoration of degraded alpine and subalpine ecosystems in the Alps National Parks, NSW, Australia', in Harmon, D. and Worboys, G. L. (eds) *Managing Mountain Protected Areas: Challenges and Responses for the 21st Century*, Colledara, Italy, Andromeda Press

Greening Australia (2004) *Easement Planting with Native Birds in Mind: High Country, Riverina, Southern Escarpment, South-West Slopes, Southern Tablelands*, Greening Australia and TransGrid, Canberra

Johnston, S. W. (1998) 'Managing degraded alpine humus soils in Kosciuszko National Park, NSW: 1. Soil properties', in *Proceedings of the ASSSI National Soils Conference*, Brisbane, Australian Society of Soil Science

Johnston, S. W. and Good, R. B. (1996) 'The impacts of exogenous zinc on the soils and plant communities of Carruthers Peak, Kosciuszko National Park', *Proceedings of ASSSI and NZSSS Conference*, vol 3, pp117–118

Johnston, S. W. and Good, R. B. (2004) 'Rehabilitating the TransGrid transmission lines in the Snowy and Brindabella Ranges', *Australasian Plant Conservation*, vol 12, p4

Johnston, F. M. and Johnston, S. W. (2003) 'Weeds set to plume following fires', *The Victorian Naturalist,* vol 120, pp194–197

Johnston, F. M. and Johnston, S. W. (2004) 'Impacts of road disturbance on soil properties and on exotic plant occurrence in subalpine areas of the Australian Alps', *Arctic, Antarctic and Alpine Research*, vol 36, pp201–207

Johnston, S. W. and Johnston, F. M. (2007) 'Vegetation succession in small gap disturbances adjacent to easements, subalpine woodland, Kosciuszko National Park, South-Eastern Australia', *Arctic, Antarctic and Alpine Research* (in press)

Johnston, S. W. and Ryan, M. (2000) 'Occurrence of arbuscular mycorrhizal fungi across a range of alpine humus soil conditions in Kosciuszko National Park, Australia', *Arctic, Antarctic and Alpine Research*, vol 32, pp255–261

Johnston, S. W., Greene, R., Banks, J. and Good, R. B. (2003) 'Function and sustainability of Australian alpine ecosystems: Studies in the tall alpine herbfield community, Kosciuszko National Park NSW, Australia', in Taylor, L., Martin, K., Hik, D. and Ryall, A. (eds) *Ecological and Earth Sciences in Mountain Areas*, Banff, The Banff Centre

14

Integrated Research on Climate Change in Mountain Ecosystems: The CLIMET Project

Daniel B. Fagre, David L. Peterson and Donald McKenzie

Introduction

Mountain ecosystems provide numerous services, such as regional freshwater supplies, but are potentially sensitive to climatic shifts because of their topographic complexity and strong environmental gradients. Like the Arctic and Antarctic regions, the higher elevations of many mountain systems have experienced relatively greater climatic change over the past several decades than lowland systems (Bradley et al, 2004). Studying climatic change impacts to relatively intact mountain protected areas offers the opportunity to better understand the mechanisms underlying ecological responses and to monitor rates of change without the confounding effects of land-use/land-cover changes that are typical of managed areas. As the regional landscapes around them change, mountain protected areas, such as national parks, become increasingly important to society as biodiversity reserves and engines of economic growth through tourism. What happens to mountain protected areas under future climate change matters more to regional human populations now than it did in the past.

Studying mountain ecosystems explicitly requires interdisciplinary and integrated approaches. A global change research programme was established in a variety of national parks of the US national park system in 1991 following an extensive competitive process that evaluated needs of multiple stakeholders from the national scale to the local park manager and community. Glacier National Park, Montana and Olympic and North Cascades National Parks, located in Washington, were among the first areas to be part of the national programme (see Figure 14.1). Research in each of these areas emphasized different ecosystem components and used different approaches to identify and understand mountain ecosystem responses to climatic variability and change to better equip managers for future decision-making. In 1998, the research programmes were merged to examine how different mountain ecosystems respond to climate change along a gradient of precipitation, landscape fragmentation and other factors from the Pacific Ocean to the Great Plains (Fagre and Peterson, 2000).

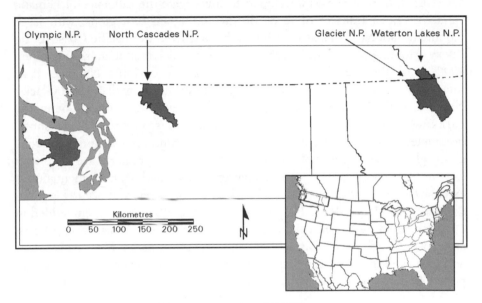

Figure 14.1 Location of CLIMET area

Note: Studies were focused on national parks along a longitudinal gradient from marine to continental climate

The Climate Landscape Interactions – Mountain Ecosystem Transect (CLIMET) uses an interdisciplinary approach to: first, document climatically-driven changes to natural resources; second, understand underlying mechanisms of change; and third, project potential future changes to mountain ecosystems. Guided by a common set of research questions and using standard protocols across all CLIMET locations, biophysical responses to climatic variability and change are explicitly identified, compared and used to inform specific management issues. We can expect non-linear responses in different regional settings with different sensitivities, with abrupt changes when thresholds are exceeded. For example, the response to a 10 per cent precipitation increase may hardly be detectable in the temperate rainforest of the western Olympic Peninsula, while the same 10 per cent precipitation increase may enhance vegetative productivity in the arid plains east of the North Cascade Range. Finally, regionally scaled assessments are relevant both to management policy for the Columbia River system that receives water from numerous mountain systems, and upon which millions of people depend, and to evaluation of future wildland fire risk, an issue that is receiving the highest priority in the CLIMET area and the American West.

Description of CLIMET area

The CLIMET project was developed to investigate the influence of climatic variability on a transect of three distinct mountain bioregions, with large mountain national parks as core research sites, from the Pacific Coast to the Rocky Mountains. CLIMET crosses three parallel mountain ranges along 1000km of the US–Canada border, centred on Olympic National Park (maritime climate), North Cascades National Park (intermediate) and Glacier National Park (continental climate). These are large, wilderness-dominated parks with minimal human disturbance and infrastructure. All are snow-dominated environments, have extensive coniferous forests, and numerous alpine glaciers and perennial snowfields. Precipitation gradients are steep with wet western aspects and dry eastern environments. Landscape fragmentation is extensive outside Olympic National Park, with forest removal occurring up to the park boundary. Glacier National Park, on the other hand, is embedded in a matrix of national forest wilderness-designated lands, Indian reservations, and, to the north, Canadian national and provincial parks. Most native flora and fauna still thrive in each national park, and hydrologic processes are unimpeded by dams or diversions.

The CLIMET integration approach

The design of the CLIMET project is built on the premise that interdisciplinary research is needed to address most issues in climate change science. This approach requires that scientists from different disciplines and with different perspectives work together in designing studies, collecting data and making interpretations. CLIMET uses the following integration approach to address climate change issues:

- develop a cadre of scientists, students and other cooperators with a common interest in the effects of climate change on natural resources;
- synthesize existing data on climate, biology and hydrology from the three CLIMET regions;
- collect data on climate, biology and hydrology in key watersheds in each of the three CLIMET regions;
- use appropriate modelling techniques to synthesize data and quantify the effects of climate change on natural resources;
- apply research results and inferences to specific resource issues by working with land managers, and by disseminating information to stakeholders.

Interdisciplinary cadre

The CLIMET research team consists of scientists from the US Geological Survey, US Department of Agriculture (USDA) Forest Service and many universities. Individuals on the team have expertise in forest ecology, paleo-ecology, hydrology and atmospheric science. Graduate students are key members of the team and provide leadership in empirical data collection. Annual meetings and frequent conference calls ensure information sharing and communication of ideas. One of the challenges of interdisciplinary work was maintaining effective communication among all cooperators. This was done through the usual means of workshops, conference calls and email, with emphasis placed on attention to the primary CLIMET objectives, consistent study protocols and meeting deadlines.

Data synthesis

Data sources were identified early in the project. We compiled all available data on climate, geology, hydrology, vegetation and ecological disturbance in the three CLIMET regions. These data were integrated within a GIS, allowing for data sharing and analysis of multiple data layers within and between regions. Data synthesis focused on two watersheds in each region – one on the west side and one on the east side – in order to represent contrasting climatic influences in wetter and drier environments. A central repository for all data-bases, as well as public posting of metadata, ensured the integrity and dissemination of project information.

New empirical data

Key data sets were collected within the watershed framework mentioned above. Specifically, we focused on the following watersheds: Glacier National Park – Lake McDonald, St Mary's Lake; North Cascades National Park – Thunder Creek, Stehekin River; and Olympic National Park – Hoh River, Dungeness River. These are all large watersheds, tens of thousands of hectares in area. Data were collected on various aspects of vegetation, ecological disturbance and hydrology in formats suitable for input to simulation models. Special emphasis was placed on developing and consistently implementing data collection protocols throughout the project. This was ensured by frequent discussions among scientists and field crew members, and by checking of data after each field operation.

Modelling framework

Ecosystem models provide insights on underlying mechanisms of ecosystem response to climatic variability and provide a look into possible futures. The Regional Hydro-ecological Simulation System (RHESSys) provides spatially explicit calculations of daily tree growth across the mountain landscape, mapped on an annual basis. Initial output from this model suggests that increased climatic variability will have as much influence on mountain forests as increases in annual average temperature or precipitation (White et al, 1998). The FIRE-BGC (BioGeoChemical) model emphasizes structural components of forest ecosystems and includes forest fires as a key disturbance factor. The model demonstrates that, in a warmer climate, forest landscapes will generate more frequent and severe fires than experienced historically even with increased precipitation (Keane et al, 1997). This nearly doubles smoke emissions in the future, reducing air quality in locations where clean air is highly valued.

Management applications

The BIOME-BGC model was recently used to examine the relative role of mountain ecosystems in providing ecological services. Cell-by-cell Pearson correlation coefficients of evapotranspiration versus precipitation, temperature and incident short-wave radiation indicate that precipitation explained spatial distributions of vegetation production and carbon stocks in vegetation and soil in drier areas, while temperature explained spatial variations in wetter areas (Fagre et al, 2005). Using BIOME-BGC and several climate change scenarios, we learned that reduced water supplies (outflows) are expected from mountain ecosystems across the region. However, the greatest effects were at mid-elevation sites, where most of the forests grow and where human settlement is increasing. Model output underscores the critical role of mountains in regional water supplies and our need to better understand ecosystem responses to climatic variability. Wildfire hazard will increase, but the amount and predictability of the regional water supply will be of even greater concern. The highest mountains where snow is stored may be even more critical in providing ecosystem services in the future.

Results of CLIMET research are also helping to inform resource management plans and other planning documentation in national parks in the region. This includes explicit consideration of climate change as a factor that affects natural resources, as well as potential means of adapting to altered environmental conditions. There is growing recognition that, from this point forward, most major decisions about natural resources must be made in the context of the effects of climate change on dynamic, non-equilibrium ecosystems.

Documenting environmental change

Glaciers

The CLIMET project has documented a variety of responses to long-term climatic change. One prominent indication of regional climatic change is glacial recession.

Figure 14.2 Paired photographs of glaciers clearly demonstrate
their dramatic rate of shrinkage throughout the CLIMET area

Note: Shepard glacier, Glacier National Park, photographed by W. C. Alden in 1913 (*top*)
and by B. Reardon in 2005 (*bottom*)

Among the most charismatic features of the three national parks, alpine glaciers offer numerous advantages for monitoring long-term changes in climate. For example, they are easily photographed and measured, and do not adapt to changing climate as plants and animals can. Responding almost solely to climatic patterns, glaciers provide a measurable signal of change within a few years to a decade.

Glacier recession at Glacier National Park reflects changes that are occurring throughout the CLIMET area and worldwide. Only 27 glaciers remain of the original 150 that existed when the park was founded in 1910. The largest glaciers cover less than 27 per cent of their original area and are expected to disappear by 2030 if current trends continue (Hall and Fagre, 2003). The area within park boundaries covered by ice and permanent snow decreased from 99km² in 1910 to 17km² by 1998 (Key et al, 2002). Surviving glaciers have thinned by hundreds of metres and may have less than 10 per cent of the ice volume that existed at the end of the Little Ice Age (see Figure 14.2).

The demise of these glaciers is driven by temperature increases of 1.6°C during the last century, three times the global average, but also is affected by timing and duration of multi-decadal droughts and pluvial periods. Temperature records have been regularly broken throughout the CLIMET study area, as warming continues (see Figure 14.3). Because the winter minimum temperatures have risen significantly, the glaciers and snowpack are warmer and initiate melting earlier in the spring, leading to a much longer melt season.

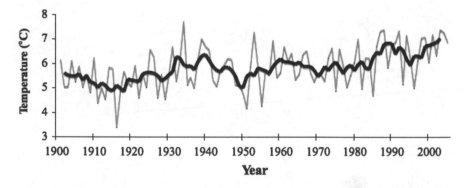

Figure 14.3 Annual mean temperature trend for western Montana, part of the CLIMET area

Note: The dark line is a five-year running average. Western Montana is currently 10°C warmer than 100 years ago

Source: Courtesy of Greg Pederson

A similar story of shrinking glaciers exists for the North Cascades National Park, which has most of the remaining glaciers in the conterminous US (Granshaw, 2002), and for the Blue Glacier in Olympic National Park

(Conway et al, 1999). Monitoring of the South Cascade glacier in Washington has served as a benchmark for regional glacier mass balances (Bidlake et al, 2005) and shows significant reductions in ice volume. Finally, Fountain (2006) has compiled data on the declining area covered by glaciers for the CLIMET region. All the glaciers reflect downward trends in size, to varying degrees.

There are numerous consequences of disappearing alpine glaciers in western mountains, particularly in the more arid areas. Glaciers provide water in late summer when other sources are depleted, ensuring the survival of aquatic organisms that require cold water (Pepin and Hauer, 2002). Many aquatic biota are narrowly adapted to specific thermal conditions and will become locally extirpated when stream water temperatures rise. Conversely, disappearing glaciers provide new habitat for colonizing plants that, in turn, attract species such as bighorn sheep (*Ovis canadensis*). Finally, there is the aesthetic impact to tourists for whom vistas of alpine glaciers are highly valued. It is unknown whether this sense of loss will have any effect on visitation rates, but park interpreters report more questions from tourists about the disappearing glaciers than in the past, and some tourists state that they are visiting now, before the glaciers are gone.

Snow

The mountain areas of CLIMET are snow-dominated environments with most (70–99 per cent, depending on the elevation) of the annual precipitation arriving as snow. Thus, snow controls many essential ecosystem functions. Mote et al (2005) have shown a 50-year decline in mountain snowpacks over the CLIMET area, which ostensibly explains the shrinking glaciers. Average snowpacks in northwestern Montana also show the 50-year decline but have not significantly changed when the entire 84 years are considered (see Figure 14.4). This lack of a snowpack trend is despite a 10 per cent net increase in annual precipitation over the past century (Selkowitz et al, 2002). Thus, the ratio of rain to snow has increased with the 1.6°C increase in mean annual temperature, somewhat diminishing the role of snow in ecosystem processes.

Snowpacks have a strong periodicity associated with the Pacific Decadal Oscillation (PDO) (Selkowitz et al, 2002), a 20–30 year cycle of anomalous sea surface temperatures that affect regional climate patterns over the CLIMET study area (Mantua et al, 1997). The result is that multidecadal periods of greater or lesser snowpacks are a prominent pattern in CLIMET mountain ecosystems (McCabe and Dettinger, 2002). Superimposed on this strong oscillation of snowpack dominance is the long-term temperature increase of the past century that has resulted in earlier initiation of snowmelt and spring runoff (Stewart et al, 2005).

Periodic summer droughts have enhanced the effect of winter snowpack variability since 1540 (Pederson et al, 2004). Coupled with the PDO effect on winter snowpack, this pattern of extreme droughts helps explain changes in

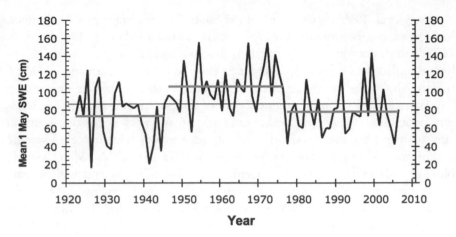

Figure 14.4 Mean snow water equivalent (SWE) of snow measured on 1 May each year
in the Many Glacier area of Glacier National Park, Montana

Note: No significant change occurs over period of record but three distinct phases are tied to the Pacific
Decadal Oscillation

Source: Courtesy of Greg Pederson

glacier mass balance, and glacier growth or shrinkage, over the past several
centuries. Because snowpack is critical to regional water supplies, the effects of
PDO and long duration droughts must be accounted for in comprehensive
water management and forecasting in the face of climatic change.

Snow avalanches

Large-magnitude snow avalanches pose hazards to humans and infrastructure,
but also shape ecosystem pattern and process by carving linear swaths out of
the forest from ridge top to valley bottom. Avalanche paths create vegetation
diversity, are utilized by wildlife, act as fuel breaks to large fires (Patten and
Knight, 1994), and transport soil and nutrients from high elevations to valley
bottoms (Butler et al, 1992). In alpine basins of Glacier Park, snow avalanches
are more important disturbance processes than forest fires. Thus, changes to
snow avalanche frequency may have profound effects on ecosystem dynamics
and the state of natural resources.

Recent increases in rockfalls, floods and debris flows in Glacier National
Park may reflect melting of permafrost in high-elevation topography and
increases in extreme rainfall events. Historically, snow avalanches have caused
extensive damage to the road bed and retaining wall of the Going-to-the-Sun
road that bisects the park. In addition, for each day that the road is closed due
to avalanches, the regional economy loses US$1.1 million (Fagre and Klasner,
2000). Thus, natural hazards linked to climatic change are of great concern to

managers and the public. Weather patterns that trigger avalanches in Glacier National Park (Reardon et al, 2004) may become more frequent in a warmer climate and should be considered in future climate change scenarios.

Forest responses

Like glaciers and snow, the alpine treeline has responded to the century-long climatic change but in more complex ways. For instance, Butler and DeChano (2001) used repeat photography to show upward treeline elevation changes, but attribute some of these shifts to changes in fire management policy. Roush et al (2007) show a variety of treeline responses to a more benign climate, but found that geomorphology was a strong interacting factor in the resulting spatial patterns. Seedling establishment at one site in Glacier National Park above the existing treeline reflected PDO phases and was driven by positive feedbacks from snow and vegetation (Alftine et al, 2003). Bekker (2005) also reported PDO periodicity in tree establishment and growth along 'fingers' of trees invading open areas where consistent high winds made establishment difficult. These cyclical processes influence the pace and pattern of tree invasion into the alpine tundra. However, infilling of gaps in the extant patch forests and krummholz was more consistent, leading to a more sharply defined treeline and greater biomass (Klasner and Fagre, 2002; Roush et al, 2007).

Regeneration of conifers in high-elevation forest ecosystems has increased throughout western North America during the past century, particularly since the 1930s (Rochefort et al, 1994). This new establishment of trees is particularly prominent near treeline during periods of low snowpack, often associated with warm PDO regimes. Subalpine meadows also have filled in with invading trees throughout the CLIMET area (Peterson, 1998; Bekker, unpublished data) (see Figure 14.5), reflecting changes in snowpack magnitude and duration as well as other climatic trends.

Peterson and Peterson (2001) ascribe the different growth patterns of trees at different elevations to periods of variable snowpack; low snowpacks facilitated tree growth at upper elevations but inhibited growth at lower treeline, with the reverse being true for large snowpacks. Finally, McKenzie et al (2004) analysed tree-growth chronologies in western North America and found that the growth rate is increasing in many high-elevation forest ecosystems below treeline. This growth increase started after 1850, concurrent with the start of industrial activity and increased greenhouse gas emissions. These increased rates of tree growth contribute to greater biomass, stored carbon and fuel for forest fires.

Cascading consequences from these tree responses to climatic change include a reduction in habitat for alpine plants and animals as forests expand. Some animals, such as pikas (*Ochotona princeps*), are also directly affected by warmer temperatures in addition to forage and habitat responses (Beever et al, 2003). Pika populations have disappeared in several areas where they used to be common.

Figure 14.5 Invasion of young trees into subalpine meadows
in Preston Park, Glacier National Park

Note: Many of the young trees are 20–50 years old and the large trees in the background are 250–300 years old. Warming temperatures and changing snowpacks have allowed the younger trees to become established during the last few decades.

Source: Daniel Fagre (2006)

Wildfire

Productivity, succession and large-scale vegetation patterns in the CLIMET area are controlled by ecological disturbance, especially fire. The area burned in any particular year is at least partially related to climatic variability, such as the PDO, and long-term climatic change. For example, years with fire area >80,000ha in national forests of Washington and Oregon are nearly four times more common during a warm PDO than during a cool PDO (Mote et al, 1999). This regional effect is moderated by synoptic-scale meteorology, particularly the effect of high-pressure ridges from eastern Washington to western Montana (Gedalof et al, 2002). Fire extent is also affected by long-term climatic change. A fire history for Glacier Park (Barrett, 1988) showed that fire frequency was lower during the height of the Little Ice Age, with a period of increased fires during the early 20th century related to warmer climate (Pederson et al, 2006).

Historical analogues help us estimate the potential for forest fires under future climate scenarios. Keane et al (1996) used FIRE-BGC to show that the resulting more-productive forest landscapes will generate more frequent and severe fires than the same landscapes experienced historically, even with the increase in annual precipitation (Keane et al, 1997). Because fire frequency has been altered by humans throughout the CLIMET area for the past century, fuel loads have built up to levels that could lead to fire intensities higher than might have been experienced in the past. A warmer climate will bring extended fire seasons and perhaps more large fires in much of the CLIMET area (McKenzie et al, 2004).

Management of forest fires is an important issue in the arid western US. In 2003, 13 per cent of the 4082km^2 of Glacier National Park was burned in three large fires and numerous smaller fires. Because forest fires are projected to become more frequent and intense under continued climate change, managers may be able to accomplish some goals, such as preserving threatened wildlife populations, by altering the management of fires. In 2003, efforts were success-fully made to divert fires away from grizzly bear (*Ursus arctos horribilis*) habitat that contained huckleberry (*Vaccinium spp.*) plants necessary to ensure bear survival through the winter. The fires were also directed away from riparian areas so that trees could continue shading streams and prevent water from increasing in temperature. Fire has been used under controlled conditions to maintain some open meadow habitats. Finally, decisions about allowing some naturally started fires to burn and others to be controlled are being made with a complex set of considerations that include the volume of smoke emissions and consequent effects on human health.

Projecting future CLIMET responses to climatic change

The CLIMET project has made considerable progress in understanding and quantifying responses of mountain ecosystems to climatic variability and has developed models to better utilize this knowledge. One such model is based on an analysis of biophysical and climatic variables to develop predictions of how dominant tree species will respond to long-term climate change. Tree data are from 10,653 forest resource inventory plots in the state of Washington, occupying a longitudinal gradient from the crest of the Cascade Range to the western slope of the Rocky Mountains. For the majority of species, it is possible to fit variables from both moisture and temperature categories of predictors that indicate optimum responses. Douglas fir (*Pseudotsuga menziesii*) trees, for instance, are predicted to be most likely to occur where growing degree days are between 2500 and 3000, and soil drought days are between 100 and 150. Climatic conditions (for example, growing degree days) can be mapped from various climate-change scenarios to show where Douglas fir trees will grow in the future. When the predictor variables are spatially explicit, then changes in the geographic niches of species in response to climatic change scenarios can be quantified. This was accomplished for 13 species, at multiple spatial scales, providing a clearer picture of possible forests of the future.

A similar approach, using an environmental envelope of optimal climatic variables for major vegetation types, mapped the distribution of vegetation for each decade into the future in Glacier National Park (Hall and Fagre, 2003). These spatially explicit projections of future vegetation neither incorporate major disturbances, such as forest fires, nor account for plant competitive processes. However, they indicate where plant species are expected to occur in a warmer climate. Of particular concern to park managers is that this scenario indicates a conversion from forest to grassland, altering habitat for animal species and affecting potential fire frequency and magnitude.

In addition, FIRE-BGC was used to examine spatial attributes of ecosystem processes rather than the more common spatial analyses performed on structural components such as cover types (Keane et al, 1999). When fire is present under both current and future climate scenarios, patch density for vegetation compositional landscape maps decreased or remained stable, meaning that forest type did not change much. However, patch density increased for net primary productivity, indicating that fire increased the spatial heterogeneity of productivity. Areas where forest became more productive were mapped to indicate shifts in dynamic ecosystem outputs. The ability to 'see' and spatially analyse processes, in addition to visible structural components of ecosystems, provides a tool for understanding drivers of biodiversity in mountains and the potential effects of climatic change on natural resources.

Conclusions

The CLIMET project has addressed issues of both scientific and management significance by integrating monitoring and research studies across the three mountain bioregions. Combining field-based studies and modelling, common processes have been investigated with comparable techniques and at similar scales. In addition to elucidating the common drivers of ecosystem dynamics and changes to mountain natural resources, the relative importance of mountains to the regional landscapes has been identified.

The perspectives provided to the public by the CLIMET project would not be possible without the interdisciplinary and integrated approaches taken with diverse partners. The public's response is evident in the frequency with which these mountain protected areas are invoked as signals of climate change in the media, community symposia and legislative initiatives. The Waterton-Glacier International Peace Park, a world heritage site that includes Glacier National Park, has even been petitioned for endangered status based on the documented effects of climate change through the CLIMET project. Managers, too, have responded by incorporating climate change considerations in resource decisions and outreach to the public.

The CLIMET approach has been broadened to encompass the Western Mountain Initiative, a group of mountain research programmes that has been continuously monitoring and investigating various mountain processes since 1981. Additionally, the Consortium for Integrated Research in Climate Sciences in Western Mountains has been formed to expand on CLIMET concepts and incorporate almost all mountain systems in the western US (www.fs.fed.us/psw/cirmount/). These efforts are providing decision-makers and a variety of stakeholders with the knowledge and tools needed to navigate the challenges of a greenhouse climate.

References

Alftine, K. J., Malanson, G. P. and Fagre, D. B. (2003) 'Feedback-driven response to multidecadal climatic variability at an alpine forest-tundra ecotone', *Physical Geography*, vol 24, pp520–535

Barrett, S. W. (1988) *Fire history of Glacier National Park: McDonald Creek Basin: Final Report*, West Glacier, Glacier National Park

Beever, E. A., Brussard, P. F. and Berger, J. (2003) 'Patterns of apparent extirpation among isolated populations of pikas (*Ochotona princeps*) in the Great Basin', *Journal of Mammalogy*, vol 84, pp37–54

Bekker, M. F. (2005) 'Positive feedback between tree establishment and patterns of subalpine forest advancement, Glacier National Park, Montana, U.S.A.', *Arctic, Antarctic, and Alpine Research*, vol 37, pp97–107

Bidlake, W. R., Josberger, E. G. and Savoca, M. E. (2005) 'Water, ice, and meteorological measurements at South Cascade Glacier, Washington, balance year 2003', *US Geological Survey Scientific Investigations Report 2005-5210*, US Geological Survey, Reston, VA

Bradley, R. S., Keimig, F. T. and Diaz, H. F. (2004) 'Projected temperature changes along the American Cordillera and the planned GCOS network', *Geophysical Research Letters*, vol 31, L16210, DOI:10.1029/2004GL020229

Butler, D. R. and DeChano, L. M. (2001) 'Environmental change in Glacier National Park, Montana: An assessment through repeat photography from fire lookouts', *Physical Geography*, vol 22, pp291–304

Butler, D. R., Malanson, G. P. and Walsh, S. J. (1992) 'Snow-avalanche paths: Conduits from the periglacial-alpine to the subalpine depositional zone', in Dixon, J. C. and Abrahams, A. D. (eds) *Periglacial Geomorphology*, London, John Wiley and Sons

Conway, H., Rasmussen, L. A. and Marshall, H. P. (1999) 'Annual mass balance of Blue Glacier, U.S.A.: 1957–97', *Geografiska Annaler*, vol 81 A, pp509–520

Fagre, D. B. and Klasner, F. L. (2000) 'Application of snow models to snow removal operations on the Going-to-the-Sun Road, Glacier National Park', in Birkeland, K. (ed.) *A Merging of Theory and Practice; Proceedings of the International Snow Science Conference*, Big Sky, Montana

Fagre, D. B. and Peterson, D. L. (2000) 'Ecosystem dynamics and disturbance in mountain wildernesses: Assessing vulnerability of natural resources to change', in McCool, S. F., Cole, D. N., Borrie, W. T. and O'Laughlin, J. (eds) *Proceedings of USDA Forest Service RMRS-P-15, Wilderness Science in a Time of Change*, vol 3, Fort Collins, Rocky Mountain Research Station

Fagre, D. B., Running, S. W., Keane, R. E. and Peterson, D. L. (2005) 'Assessing climate change effects on mountain ecosystems using integrated models: A case study', in Huber, U. M., Bugmann, H. K. and Reasoner, M. A. (eds) *Global Change and Mountain Regions: A State of Knowledge Overview*, Berlin, Springer

Fountain, A. (2006) 'Glaciers Online: Glaciers of the American West', http://glaciers.research.pdx.edu/ accessed 6 November 2006

Gedalof, Z., Mantua, N. J. and Peterson, D. L. (2002) 'A multi-century perspective of variability in the Pacific Decadal Oscillation: New insights from tree rings and coral', *Geophysical Research Letters*, vol 29, pp2204–2207

Granshaw, F. D. (2002) *Glacier Change in the North Cascade National Park Complex, Washington State USA, 1958 to 1998*, Masters thesis in geology, Portland, Portland State University

Hall, M. P. and Fagre, D. B. (2003) 'Modeled climate-induced glacier change in Glacier National Park, 1850–2100', *Bioscience*, vol 53, pp131–140

Keane, R. E., Ryan, K. C. and Running, S.W. (1996) 'Simulating effects of fire on northern Rocky Mountain landscapes with the ecological process model FIRE-BGC', *Tree Physiology*, vol 16, pp319–331

Keane, R. E., Hardy, C. C., Ryan, K. C. and Finney, M. A. (1997) 'Simulating effects of fire on gaseous emissions and atmospheric carbon fluxes from coniferous forest landscapes', *World Resource Review*, vol 9, pp177–205

Keane, R. E., Morgan, P. and White, J. D. (1999) 'Temporal patterns of ecosystem processes on simulated landscapes in Glacier National Park, Montana, USA', *Landscape Ecology*, vol 14, pp311–329

Key, C. H., Fagre, D. B. and Menicke, R. K. (2002) 'Glacier retreat in Glacier National Park, Montana', in Williams, R. S. Jr. and Ferrigno, J. G. (eds) *Satellite Image Atlas of Glaciers of the World, Glaciers of North America: Glaciers of the Western United States*, Washington DC, United States Government Printing Office

Klasner, F. L. and Fagre, D. B. (2002) 'A half century of change in alpine treeline patterns at Glacier National Park, Montana, U.S.A.', *Arctic, Antarctic, and Alpine Research*, vol 34, pp49–56

Mantua, N. J., Hare, S. R., Zhang, Y., Wallace, J. M. and Francis, R. C. (1997) 'A Pacific interdecadal climate oscillation with impacts on salmon production', *Bulletin of the American Meteorological Society*, vol 78, pp1069–1079

McCabe, G. J. and Dettinger, M. D. (2002) 'Primary modes and predictability of year-to-year snowpack variations in the western United States from teleconnections with Pacific Ocean climate', *Journal of Hydrometeorology*, vol 3, pp13–25

McKenzie, D., Gedalof, Z., Peterson, D. L. and Mote, P. (2004) 'Climatic change, wildfire, and conservation', *Conservation Biology*, vol 18, pp890–902

Mote, P. W., Keeton, W. S. and Franklin, J. F. (1999) 'Decadal variations in forest fire activity in the Pacific Northwest', *11th Conference on Applied Climatology, American Meteorological Society*, pp155–156

Mote, P., Hamlet, A., Clark, M. and Lettenmaier, D. (2005) 'Declining mountain snowpack in western North America', *Bulletin of the American Meteorological Society*, vol 86-1-39, DOI: 10.1175

Patten, R. S. and Knight, D. H. (1994) 'Snow avalanches and vegetation pattern in Cascade Canyon, Grand Teton National Park, Wyoming, U.S.A.', *Arctic and Alpine Research*, vol 26, pp35–51

Pederson, G. T., Fagre, D. B., Gray, S. T. and Graumlich, L. J. (2004) 'Decadal-scale climate drivers for glacial dynamics in Glacier National Park, Montana, USA', *Geophysical Research Letters*, vol 31, L12203, DOI:10.1029/2004GL0197770

Pederson, G. T., Gray, S. T., Fagre, D. B. and Graumlich, L. J. (2006) 'Long-duration drought variability and impacts on ecosystem services: A case study from Glacier National Park, Montana USA', *Earth Interactions*, vol 10, pp1–28

Pepin, D. M. and Hauer, F. R. (2002) 'Benthic responses to groundwater–surface water exchange in two alluvial rivers', *Journal of the North American Benthological Society*, vol 21, pp370–383

Peterson, D. L. (1998) 'Climate, limiting factors, and environmental change in high-altitude forests of western North America', in Beniston, M. and Innes, J. L.(eds) *The Impacts of Climate Variability on Forests*, Heidelberg, Springer

Peterson, D. W. and Peterson, D. L. (2001) 'Mountain hemlock growth responds to climatic variability at annual and decadal scales', *Ecology*, vol 82, pp3330–3345

Reardon, B. A, Fagre, D. B. and Steiner, R. W. (2004) 'Natural avalanches and transportation: A case study from Glacier National Park, Montana, USA', in *Proceedings of the International Snow Science Workshop*, 19–24 September, Jackson, Wyoming

Rochefort, R. M., Little, R. L., Woodward, A. and Peterson, D. L. (1994) 'Changes in the distribution of subalpine conifers in western North America: A review of climate and other factors', *Holocene*, vol 4, pp89–100

Roush, W. J., Munroe, J. S. and Fagre, D. B. (in press) 'Development of a spatial analysis method using ground-based repeat photography to detect changes in the alpine treeline ecotone, Glacier National Park, Montana', *Arctic, Antarctic, and Alpine Research*, vol 39, pp297–308

Selkowitz, D. J., Fagre, D. B. and Reardon, B. A. (2002) 'Interannual variations in snowpack in the Crown of the Continent Ecosystem', *Hydrological Processes*, vol 16, pp3651–3665

Stewart, I., Cayan, D. and Dettinger, M. D. (2005) 'Changes towards earlier streamflow timing across western North America', *Journal of Climate*, vol 18, pp1136–1155

White, J. D., Running, S. W., Thornton. P. E., Keane, R. E., Ryan, K. C., Fagre, D. B. and Key, C. H. (1998) 'Assessing simulated ecosystem processes for climate variability research at Glacier National Park, USA', *Ecological Applications*, vol 8, pp805–823

15

Integrated Approaches to Research and Management in Mountain Areas: Synthesis and Lessons for the Future

Paul Mitchell-Banks and Martin F. Price

The words 'integrated approaches to research and management in mountain areas' give a hint of the challenges that these approaches pose, and the opportunities they present, to government or private land managers, researchers and other stakeholders. Integration of any kind poses challenges, and mountains are particularly challenging systems to access, work within, and understand. The introductory chapter described how this book is built around three major themes: research beyond disciplinary boundaries; integrated approaches to managing ecosystems; and mountain environments and the people who depend on them. This final chapter synthesizes the key issues that emerge from the different chapters and identifies key lessons for undertaking integrated research and management.

Organizing principles

This review and synthesis is primarily organized around the 11 principles of research partnership, first developed in 1998 by the Swiss Commission for Research Partnership with Developing Countries (KFPE, 2003) and subsequently applied in many countries in different contexts. These linked principles recognize the need for mutual respect, honesty and openness in a partnership that is characterized by effective communication and a commitment to a long-term relationship. Relationships require hard work from all stakeholders to survive and thrive – and the establishment of mutual trust requires ongoing dialogue, sharing experiences, mutual learning and the involvement of all those with a stake. The hard work that all stakeholders have to contribute to successfully undertaking integrated research and/or management is not just limited to developing countries, but indeed is required anywhere on Earth. The 11 principles are:

1   Decide on objectives together.
2   Build up mutual trust.
3   Share information and develop networks.
4   Share responsibility.
5   Create transparency.
6   Monitor and evaluate collaboration.
7   Disseminate the results.
8   Apply the results.
9   Share profits equitably.
10 Increase research capacity.
11 Build on the achievements.

These principles provide a good foundation on which to structure the review and synthesis of this book's chapters, particularly because of its focus on mountain areas. Integrated research and management in these areas requires extra attention to making any relationships work, not only because of the complex physical geography that can create difficulties for access and communication – often resulting in longer time-frames and higher costs – but also because of the need to deal with particularly complicated ecological, sociological and economic milieus – often both within the mountain areas themselves and in other areas to which mountains provide key services.

We suggest that two additional principles should be added to the 11 principles presented above: first, incorporate traditional ecological knowledge and/or local knowledge; and second, integrate legislation and policy. Also, given the focus of this book on not only research, but also management, we have revised principle 10 by deleting the word 'research'. Below, we discuss each of the 13 principles briefly before presenting a review and final synthesis of the previous chapters of this book in the context of the principles.

Decide on objectives together

Ideally, a research project or management programme should be discussed with all those involved, as well as those both directly and indirectly impacted or affected. Inclusion of all stakeholders from the earliest possible stage, ideally starting with the preliminary planning, helps to contribute to a more coherent definition and understanding of the research or management activities – and this facilitates more informed participation and cooperation and the establishment of a 'team' mentality based on shared objectives.

Build up mutual trust

Cooperation, especially in a challenging research and/or management project, can only flourish in an atmosphere of trust – which requires time and patience, and the ability to be empathetic about the positions of all members of a team and those affected directly or indirectly by the work being done. Building trust requires hard work on the part of all, and a willingness to set aside prejudices or beliefs and to see things, places and people in a fresh and welcoming light. Historically positive interactions promote trust, and often researchers will look to work with old and 'trusted' partners – but there is also a need to systematically look for and develop new contacts and relationships to sustain the community in which the research or management is being undertaken. This new recruitment of partners not only fills the positions of those who move, change jobs, or fall ill or die, but also leads to 'fresh' eyes looking at the challenges and bringing in new experiences, education and awareness to address the problems being faced.

Share information and develop networks

Ongoing and effective communication plays a key role in facilitating the co-operation that is key to effective integrated research and management. The sharing of information needs to be sensitive to the challenges and opportunities that all partners face in communicating not only with each other, but with the world in general. Networks survive and thrive on effective communication. Every integrated research or management project should have a communications strategy that directly identifies the information needs of each audience, including those assessing the information, and is also sensitive to their technological and infrastructural limitations.

Share responsibility

In principle, responsibility should be shared by all partners at all levels of the project, as this creates a sense of involvement and belonging to a team. This can help prevent the sense of 'someone else' having all the control, and also create opportunities for greater input through the introduction of new perspectives and for less experienced partners to develop and advance their research and management skills and experience.

Create transparency

Ideally, there should be transparency or equality in how the contributions of partners are considered, whether these are monetary or contributions in kind. Equally, all partners should have a role in making decisions on the allocation of resources and staff and, as much as possible, the 'weight' of their vote should not be tied directly to the level of their contributions.

Monitor and evaluate collaboration

The collaboration or partnership needs to be constantly monitored and evaluated, and assessments made on a regular basis. The assessments need to cover a wide range of issues, including management, communication, decision-making, implementation and capacity development. Monitoring and evaluation require regular information exchange and meetings and a commitment to address matters as they arise.

Disseminate the results

Access to the results and outcomes of a project should be open, and all partners in a research project should have opportunities to participate in disseminating these results and outcomes. This principle also holds for integrated management. Dissemination should be not only to the research community and/or managers but, at least as importantly, to the lay people, government staff, or business people who can use the results to improve their own lives or work performance.

Apply the results

The residents of the area in which a research or management activity was conducted should be able to apply any useful results, thus benefiting directly from the activity. This requires facilitation, for example, through extension or farmer-to-farmer activities, which goes beyond mere dissemination of results. Too often, when innovative activities are undertaken in an area, and even when there has been an attempt made to disseminate the results, any lessons or benefits are essentially 'lost' to the area because of a lack or loss of champions or stewards to ensure that these benefits are applied.

Share profits equitably

The intellectual and potential commercial value of any research results have to be shared fairly with all the partners, and all partners should be credited for their contributions to the research.

Increase capacity

Both research and management should contribute to institutional and individual growth. This requires a thorough understanding of the strengths and weaknesses of the various partners and how they can best support each other to increase capacity, not only for research, but also for management.

Build on the achievements

The legacy from any research or management project should involve not only new knowledge and a contribution to sustainable development but also improved research and management capacities. Thus, knowledge and understanding has to be recorded, made easily accessible, and then applied. This legacy requires that the involved institutions continue to operate and that the scientists involved continue in employment – or, if this is not possible, that their knowledge is recorded; corporate memory is often lost through job losses, moves to new employment, and the closure/amalgamation of institutions. Building on the achievements means thinking long-term, and anticipating how best to keep the lessons of earlier work accessible, understood and applied in current research and management.

Incorporate traditional ecological knowledge and/or local knowledge

A key element of integrated research and management is to incorporate traditional ecological knowledge (TEK), or the local knowledge of the people of the area, within the project design and to include it appropriately when applying conventional scientific methods. This principle has been added to emphasize that promoting participation from local people (as suggested in the first principle of deciding on the objectives together) does not always explicitly support the incorporation of TEK or local knowledge in the design of research and/or management projects.

Integrate legislation and policy

Integrated research can identify shortfalls in existing policy and legislation and, at times, suggest ways in which policy and legislation can be amended to avoid or minimize the creation of the problems resulting from these shortfalls. At times, integrated research will address a legacy problem, or suite of problems, deriving from poorly designed and/or implemented legislation and policy; at other times, it can identify future problems that current legislation or policy will not effectively address. In short, integrated research should identify and support integrated legislation and policy, and their effective implementation through informed management.

Review and synthesis of the chapters

The chapter by Messerli and Messerli, addressing the development of interdisciplinarity and transdisciplinarity, begins with a wonderful 200-year-old quote from Alexander von Humboldt: 'While each series of facts needs to be studied separately in order to discover its own rules of order, the general study of nature demands all knowledges ... be then combined.' This is the very heart of the idea of integrated research – and its greatest challenge. Not only is there the challenge of achieving understanding within the individual disciplines and research areas – but there is the additional challenge of working effectively together, to see how the sum of the whole is more than the sum of the parts; in effect, how the general study of nature is organismic and not mechanistic. Integrated research is required to appreciate the organismic nature of this world.

The Swiss Man and Biosphere Programme in the Swiss Alps led to recognition of the subtle balance between tourism, agriculture and the environment that had developed over time, and that this was based on socio-economic structures that were not sustainable over the long term. Subsequent research involved a participatory process (involving scientists, residents and government officials) to formulate more sustainable policies and to enact long-term goals that effectively addressed the environmental, socio-cultural and economic aspects of sustainability. The MAB initiative demonstrated the importance of adequate funding to establish strong leadership from a multi-disciplinary and committed expert group that was committed to excellence, open to interdisciplinary approaches, and had excellent programme management and qualified scientists. Key elements of this interdisciplinary initiative can be found in the transdisciplinary work that is done in much of the developing world, incorporating greater roles for stakeholders and practitioners in order to address the important role of societal inputs into environmental (and often the resulting economic) conditions. The key role of public participation and local civil institutions in working closely together throughout the research programme remains true to this day and for the future. This integrated involvement of scientists and the public is also a key aspect of successful international research programmes. Transdisciplinarity captures the idea of different academic disciplines working in close cooperation with practitioners and stakeholders, in order to explicitly marry scientific and socio-economic challenges and opportunities to more effectively address key problems and symptoms.

This first chapter sets the scene for subsequent chapters not only because it emphasizes a number of key issues regarding integrated research, but also because of the long history of the research conducted and the innovative approaches that it incorporated, and the profound impact of the MAB Programme on international research. The recognition of the interconnectedness of society, the environment and the economy, and the need for specialists and stakeholders involved in all three areas to address the challenges together, are fundamental principles for integrated research in mountain areas.

Ramakrishnan's chapter on *Jhum*-centred land use in northeast India further explores this integration of science and society, arguing that many agroecosystem typologies have sub-typologies that are society-specific and adapted to diverse ecological conditions. This leads to the recognition of the cultural landscapes of each ethnic society, deriving from their perceptions and traditional value systems. These landscapes are based in TEK and as such, can only be investigated by ensuring community participation that offers an opportunity for meaningful community involvement and sharing in an environment of trust and respect. The TEK has to be effectively linked or integrated with conventional or formal science in order to understand the local situation and local landscape. Protected ecosystem areas, such as sacred groves, often contain species with keystone values within the ecosystem and, by scientifically recognizing these roles and explicitly incorporating these keystone species into land management regimes, scientists are able to not only engage the local residents but also effectively implement ecological management and research. TEK determines the integration of human activity within the landscape, as well as the outcomes of the impacts of the residents and the form and function of the landscape that sustains them.

Ramakrishnan's chapter powerfully describes issues relating to place, time and people, and shows how traditional or local knowledge can play a critical role in any integrated research project involving the environment. It shows that one cannot consider the environment without also considering the society and the economy that are directly related to and with that environment. Integrated research involves thinking beyond a single discipline or a single perspective, and the incorporation of traditional stakeholders and their remarkable body of knowledge about their environment, society and economy – and their linkages – greatly expands not only the base level of knowledge but also opens up new understandings about potential research and relationships to be explored.

The chapter by Mattson and Merrill argues that the traditional approaches to conservation design reflect a 'generalized' approach that often fails to address and engage the specific challenges and opportunities found in the specific context (social, economic and environmental) of the area of concern. Social and decision processes have to address human values, world-views and myths, as these largely shape what people expect, desire and demand from the world around them. Furthermore, the survival of democratic governments depends on the policy processes addressing the common interests over those of the minority.

Healthy human societies and healthy ecosystems are interdependent, and effectively addressing this interdependence relies on governance that focuses on commonality over individual needs, and often on taking an adaptive management approach that incorporates lessons and knowledge as they develop. The challenge here lies in effective learning, incorporation of knowledge, and action by government or other agencies – and typically this is not well done. Even where agencies are responsive and adaptable to evolving situations, there is still the challenge of political or policy influence on not only how, but also when, they act or react. In most current conservation design, the approach to problem-solving and the design of reserve or protected area networks remains tool-oriented but, in many ways, this fails to address the root cause of the problem: human beings. The essence of conservation is to stop destructive human behaviours and/or promote and preserve human actions that are beneficial. This is challenging, especially when one considers the limited resources and the difficulties of attaining sufficient insight into ecological and policy processes to allow the optimal allocation of resources across space and time.

Successful conservation depends on individual people deciding to create policies or engage in behaviours that support rather than degrade sustainable ecosystems. However, while we might think that we are a rational species, we behave – more often than not – with bounded rationality, and apply approaches that do not take into consideration the specific contextual conditions. In other words, we essentially react and do, rather than analyse and act. How people identify themselves, what expectations they hold, what demands they have, what values they hold, what strategies they employ to interact with others, and how this is seen by not only themselves but also others – these all interact with the biophysical realm to create the context in which the conservation programmes and activities have to be designed and implemented. Mattson and Merrill argue that the effective implementation of policy-oriented conservation design (POCD) requires an 'intelligence exercise' – similar to the military sense of the word – in order to apply information to achieve success.

This gathering of information is critical to deriving a body of information and understanding that is comprehensive, relevant and reliable enough to apply to resolving challenges faced in a timely and economical fashion – and is also context-specific; what is appropriate for one context may not be appropriate for another. The approach used to derive this body of information and understanding should be consistent, with a focus on decision and policy processes. Prescriptions deriving from POCD are therefore very contextual, as a result of the use of a consistent approach to analyse the specific challenges and opportunities present within a specific spatial and temporal setting. POCD should ideally consist of five main elements: first, clarification of the problem-solver's viewpoint; second, orientation to the 'problem'; third, 'mapping' or outlining the context; fourth, specification of a design, policy or approach; and fifth, subsequent appraisal and redesign. This process essentially addresses the 'who, where, why, when, what and how' aspects in deriving the appropriate policy approach.

The chapter raises the key issue of the role of governance overlying and at times even underpinning the research problems being addressed, and emphasizes the need not only for an integrated approach of disciplines/perspectives and partners (either individuals or institutions) but also the ultimate role that government agencies can play. Integrated research that is dynamic and effective can have its results severely curtailed by a government bureaucracy that is stagnant or slow to evolve and does not incorporate, on an ongoing basis, the outcomes of up-to-date and reliable integrated research. The chapter alludes to the importance of scientists being aware of how governments work (or not) and to the challenges that might be faced in having research, particularly integrated research, acknowledged and incorporated into policy and legislation.

Brown and Schreier's chapter on innovative watershed management focuses on moving away from the traditional 'engineered' approach to more innovative strategies of working with natural systems to address water management requirements. Engineered approaches involving infrastructure to retain, store, deliver and drain water as needed have been locally successful, but have not effectively addressed the needs of the watershed as a whole; infrastructure has been constructed piecemeal, with inadequate consideration for cumulative effects. Land-use intensification, growing industrial demand, urban expansion and climatic extremes resulting from global warming are all increasing the demands on watershed management and the existing infrastructure. Growing concerns about water source-to-tap protection, how much water is required to maintain aquatic biota and environmental services, and potential strategies to address increased climatic variability present further challenges. At the global scale, mountains play central roles in these issues; as 'water towers', they provide 60 per cent of global freshwater resources from less than 35 per cent of the land base.

One possible strategy is to take an adaptive and integrated approach, including the best of 'hard path' engineered approaches and placing increased emphasis on 'soft path' approaches such as education, behaviour change and environmental engineering. Simple environmental engineered solutions, such as increasing the area of porous or pervious surfaces and enlarged riparian buffer zones along streams, can have a profound positive impact on water management. Creating wetlands to act as buffers to water movement or as settling or filtration systems can greatly increase the management of water quality and flow rates. Managing 'green water' – that proportion of precipitation that is intercepted by vegetation or enters the soil and is converted into biomass and later evapotranspired – is another effective 'natural' water management strategy.

While effective land use and management is one element of a successful strategy, it focuses only on the supply side, and there is also a need to manage on the demand side. One difficulty that water management has traditionally faced is that water has been regarded as a common good, and this has resulted in it being highly undervalued. Simple strategies could reduce demand between 30–50 per cent. In short, effective water management requires an integrated approach tailored to the specific context of the challenges and opportunities of the water supply and demand. This approach needs to incorporate strategies of effective land use and management, the appropriate use of engineered infrastructure married with natural structures, and also the education and awareness and use of appropriate technology by those who access water resources.

The chapter effectively raises the issues of networks of places and people, the challenges we collectively face, and the approaches we need to collectively employ. Water for urban residents largely comes from rural sources, often in the mountains, and water management needs to be addressed at both the source and the demand side, using both 'hard' and 'soft' approaches. In many ways, these are issues about balance, and to find the right balance requires constant monitoring, linked to adaptive and/or mitigative action. Integrated research constantly has to monitor, evaluate, disseminate, increase research capacity and build on earlier achievements.

Mowo, Shemdoe and Stroud's chapter on the African Highlands Initiative and interdisciplinary research and management turns the focus to Africa and a more applied and grass-roots scenario. The overall goal of the AHI is to improve the ability of grass-roots, service and support organizations, and to contribute to policy development, in order to achieve better food and nutritional security, increase income derived from agricultural activities and improve natural resource management in the intensively cultivated and densely populated highlands of Eastern and Central Africa. The approach involves inter-institutional and multidisciplinary research and development efforts with strong community participation.

The team encountered a number of challenges to interdisciplinary work, including:

- lack of mutual understanding, respect and appreciation due to varying interests, ambitions and expectations among professionals from different backgrounds;
- limited team spirit, evidenced by the reluctance of some professionals to work together, due probably to lack of confidence, difficulty for some professionals to change from previous ways of doing things, and a reward system that does not recognize individual efforts in a team;
- logistical constraints, mainly in coordinating professionals from different institutes who have other research duties to attend to, but also because of high travel and subsistence costs, given that some AHI sites are far from the institutes where participants are based.

Once these challenges have been recognized, the successful implementation of integrated research in and for natural resource management requires:

- identification of strategic partners;
- building an interdisciplinary team spirit;
- understanding farmers' needs and ensuring that project goals meet these needs;
- building a culture of interdisciplinarity where team members are patient with each other and there is mutual learning between professions, so that perspectives are understood and shared;
- building commitment, trust and a learning culture among team members to learn the farmers' indigenous knowledge and to understand and respect their point of view;
- changing the reward system – in which, in most countries, the recognition of effort is based on individual rather than collective achievement – to promote integrated research.

Many of these strategies are common to those raised in other chapters, with the exception of those addressing rewards and the promotion of integrated research. In effect, the reward system continues to reflect the single discipline mentality and what could be considered a 'tame' set of scientific challenges rather than the complex and contentious 'wild' set of scientific challenges that are becoming increasingly obvious.

The chapter strongly shows the need to recognize that integrated research is not just research, but in fact an approach that goes beyond a single discipline mentality or a cluster of individual disciplines and moves into research that involves a spectrum of disciplines and the tools, techniques, technology and knowledge that are unique to each but that have to support each other. This is a significant challenge, not only to promote integrated research through current funding mechanisms, but also to recognize the value and contributions of the spectrum of disciplines in integrated research findings and dissemination.

Kaihura's chapter on interdisciplinary research and management in mountainous northern Tanzania discusses the importance of working closely with local stakeholders from the earliest possible stage of the project or programme, preferably from when issues, priorities and activities are defined. Ideally, researchers and scientists should accommodate the desires and needs of the local residents and stakeholders in the design and execution of the research, and there should be an atmosphere of respect and appreciation between the stakeholders and their practical/economic approach and the scientists who tend to focus more on research and scientific methods. There is an ongoing need to maintain communication and promote understanding and awareness between the local stakeholders and the researchers; they should comprehend each others' perspectives and appreciate the complementarities of their knowledge, but also be aware that this knowledge may take different forms and approaches.

At the local level, the project had many results and outcomes, including: improved on-farm biodiversity; improved agricultural intensification; increased yields and reduced soil erosion; improved nutritional value and livelihood of smallholder farmers; renewal of degraded pastures and forests that increased honey production; effective and improved land utilization; establishment of by-laws; women's empowerment; inclusion of environmental studies in the primary school curriculum; and the establishment of good working relations and farmer networks. The dissemination of these results and outcomes required various groups of stakeholders to take on different roles. Farmers made both formal and informal visits to other farmers to discuss the performance of different technologies, and exchange plant material of various crop varieties, medicinal herbs, and hedgerow and conservation structures. Farmers' groups met with each other and exchanged information and experiences, and expert farmers participated in national, district and local agricultural shows and international workshops and conferences. Political and religious leaders at the local and district levels raised awareness about the project as a whole, government staff participated in workshops, and researchers and extension staff produced pamphlets and facilitated farmers' participation in events – in short, everyone participated in the dissemination process. The media also played a key role, broadcasting workshops on both TV and radio, and projects were reported on in newspapers, articles and newsletters.

Kaihura's chapter emphasizes not only the team work required to establish and undertake integrated research projects but also the teamwork necessary to effectively disseminate and apply the results, share profits/results equally and increase research capacity. The team is a key concept or element of integrated research.

Willebrand's chapter on the Swedish Mountain Mistra programme describes a strategic approach to policy development and implementation. The programme has focused on the mountains of northwestern Sweden, an area facing a declining economy and a falling population. The traditional economic activities of forestry and dam building have declined, tourism has been growing, and the influence of non-residents on conservation agendas has been increasingly felt. The local people tend to focus on the use of natural resources, and this can lead to conflict with the conservation agenda of the non-residents.

The intent of the programme has been to develop scientifically based strategies for the management and long-term development of the resources of the mountain region; a key principle is to ensure knowledge exchange and interaction between various people within the area and involved in the programme, recognizing that information is knowledge and power. Early stages of the programme indicated that communication between stakeholders and researchers played a key role in successful implementation, but also led to a decision to decrease the number of ecological researchers and introduce economists and political scientists in the final phase. The importance of communication was also underlined by the employment of a full-time communicator with strong experience from the media field, to focus on communication both within the research team and between the stakeholders and researchers. Regular meetings played a key role in developing two-way trust between stakeholders and researchers. The final phase focused more on interdisciplinary approaches with increased interaction across disciplines, rather than the earlier approach with research from many disciplines, but little interaction or integration. The interdisciplinary approach recognized that the group had to undertake a number of research tasks collectively, rather than individually. This integrated approach, involving teamwork of people across disciplines proved to be successful, though more demanding in terms of team management.

Willebrand's chapter links strongly to Kaihura's, as clear communication is a key element for any team to function, not only within the team itself but also between the team and the various stakeholders with whom they need to effectively interact. Good integrated research requires good communication in deciding on the appropriate objectives, sharing information and developing networks, disseminating the results, and increasing research capacity.

Mitchley, Tzanopoulos and Cooper's chapter addresses the challenge of reconciling the conservation of biodiversity with declining agricultural use in the mountains of Europe. Farming practices have traditionally heavily modified the higher elevations in many mountain regions and, as a result of a period of agricultural adjustment reflecting both external subsidies and socio-economic change, many of these farming areas have experienced decline and even abandonment, resulting in landscape closure and increasing forest dominance. This return of the forest can result in the loss of traditional landscapes, but can positively contribute to the recovery of pre-agricultural ecosystems, and the reintroduction of large predators.

BioScene addressed the policy context of the 25-year period and also explored four scenarios over this period: first, Business as Usual; second, Agricultural Liberalization; third, Managed Change for Biodiversity; and fourth, Wilding. The project used an interdisciplinary approach to understand these four complex scenarios, and in doing so, employed the following strategies to overcome some of the challenges of interdisciplinary research:

* building a consortium of researchers with a positive attitude towards inter-disciplinary research and with experience working across disciplines;
* scheduling regular meetings to encourage frequent interaction between social and natural scientists;
* developing distinct disciplinary work packages to facilitate the production of disciplinary publications and other outcomes that are traditionally rewarded by the academic system – thus providing extra motivation for researchers to participate;
* using a range of methodological tools, such as scenarios, visualizations and sustainability assessment, to encourage the integration of the ecological and socio-economic strands of the research project;
* fostering a climate of mutual academic tolerance and flexibility, and allowing adaptive management of the project's work structure while still focusing on the overall aims and objectives of the research.

The chapter recognizes the important role that individual disciplines can play in integrated research and the need to combine the strengths of the individual disciplines to achieve a stronger integrated disciplinary result. The chapter also recognizes the important role that stakeholders can play as co-participants in an integrated research process – much as the two Tanzanian cases did. In each of the six BioScene study areas, there was a stakeholder panel composed of approximately 12 local people with a broad range of views, in order to promote deliberation among the stakeholders that reflected many perspectives. There was a conscious effort to have an appropriate gender and age mix, given the project's focus on past and future processes of change.

Quinn, Greenaway and Duke's chapter details how cumulative effects are insidious, given that, while each individual action, when considered in isolation, may seem insignificant, the combination of spatial and temporal interactions results in highly complex and often unpredictable changes to ecological and social systems – as seen with climate change, loss of biodiversity and habitat fragmentation. These impacts, effects or problems can be referred to as 'wicked', as they do not have a simple problem definition, and often there is considerable disagreement between stakeholders as to what the problem is. The conventional linear and causal chain approaches used in the planning and management of 'tame' problems (potentially complicated problems that can be clearly articulated and solved using the appropriate and standard methods) simply do not work.

The traditional command-and-control approach to natural resource management often hides the true nature of wicked problems and can exacerbate the cumulative effects. This can stem from various causes including, but not limited to, jurisdictional fragmentation, the piecemeal regulation of individual resource sectors, and a focus on the mitigation of local short-term effects. The addressing of a multitude of small approval challenges (subdivisions, cutblocks, recreational trails and so on) can collectively add up and create a 'tyranny of small decisions'. Sustainable development, adaptive ecosystem-based management, and other holistic approaches attempt to address the smaller decisions collectively and in an integrated approach, so that the interrelationships between the smaller issues are understood and addressed more effectively. These approaches need to be both more interdisciplinary and more integrated than historical resource management and planning. One such approach is cumulative environmental affects management. A survey of Canadian and American public land managers collectively responsible for the Crown of the Continent Ecosystem – part of the Rocky Mountains shared between the Canadian provinces of British Columbia and Alberta and the US state of Montana – showed that the most important characteristics of effective collaborative cumulative effects assessment (CEA) initiatives were:

- clearly defined and shared goals and objectives;
- shared commitment for long-term involvement;
- adequate commitment of resources;
- common issues and pressing need for response;
- frequent and effective communication;
- mutual respect and trust among participants.

The respondents to the survey also identified the most important barriers to effective collaborative resource management initiatives as:

- lack of resources, shared agency mandates and philosophies, and agency support;
- interagency barriers and 'turf' or jurisdictional issues;
- lack of continuity of participating members.

There is remarkable consistency between the success factors and barriers to effective collaborative CEA and, again, the critical role of effective communication and an integrated/collaborative approach is evident. When multi-jurisdictional cumulative effects scenarios are considered, particularly between countries, the challenges become particularly difficult. There are additional challenges in maintaining long-term research, planning and management approaches that are costly in terms of staff and resources – and for which it is often difficult to determine what progress is being made and how much additional time and staffing is required. Complex and hard-to-manage projects or programmes are difficult to continually address, given tight budgets, shifting priorities, changing governments or management, and limited staffing resources.

Quinn, Greenaway and Duke's chapter makes a number of recommendations on how best to approach CEA, particularly large-scale CEA within the multi-jurisdictional setting that is increasingly common with the growth of industrial development, populations and global trade. These recommendations, in no particular order, include: higher-level support; clear articulation of project goals and objectives; a shared approach to dedicated resource; the development of internal and external communication products; monitoring, feedback and continuous improvement; and explicit incorporation of results into agency activities. The conclusions include the following:

- Global lessons – citizens play a critical role in bringing situational information to decision-makers that is then incorporated into policy, planning and management processes; managers can not effectively manage without this bottom-up input and support. At times, the indirect benefits of greater cooperation and participation may actually outweigh the specific benefits of a cooperative approach for that specific problem, as new and/or stronger relationships and understandings are developed and future collaboration promoted.
- Operational lessons – strong leadership is required to promote and support multi-agency cooperation, through:
  - having a lead agency, though this can create suspicion among other participants;
  - having a neutral third party serve as facilitator of the project, though this requires a clear mandate and adequate financial support;
  - having a multi-agency steering committee with shared responsibility for chairing regular meetings.
  Regardless of the approach, strong leaders and 'champions' are vital; these figures or groups have to take a long-term view, with operations having to remain flexible to respond to changing budgets, election timetables, personnel changes and workloads. This demonstrates the unique roles of not only individuals but also the group.
- Long-term commitment by participants is essential, as there are no 'quick'

fixes, and participants need to focus on problems that transcend boundaries and are faced by all. The development of a common language to ensure understanding and participation by all partners and the use of consistent data protocols are both fundamental and critical to long-term success.

• Lessons relating to institutional frameworks – existing short-term planning cycles of political terms and institutional structures do not support strategic approaches. People in the field are likely to recognize the importance of integration required for collaborative CEA, and the processes they develop may well catalyse larger system changes in addressing wicked problems.

This chapter clearly recognizes the roles and responsibilities of not only the individual and the institution but also the government in integrated research. This has direct implications for not only the sharing and use of information but also its incorporation into policy and legislation.

Mitchell-Banks' chapter on the Muskwa-Kechika Management Area describes an ambitious attempt further north in the Canadian Rocky Mountains to protect wilderness and wildlife values, through establishing protected areas and special management zones in which controlled economic development was not only permitted but encouraged. This 64,000km² area is home to some of the greatest global boreal wildlife populations, but is also an area with huge oil and gas reserves, mineral resources and timber, and significant opportunities for commercial and personal recreation. The M-KMA was established under unique legislation and was meant to serve as a land-use planning and management model, and as a new way to plan more effectively and manage sustainably.

The M-KMA is effectively the 'offspring' of three land and resource management plans, stakeholder-driven processes in which the government provides a supporting role and that result in a plan that is then sent to the government for final approval, amendment or rejection. The M-KMA is subject to five unique land-use plans that complement existing government planning requirements and effectively raise land-use planning and management to a higher level than for most of the province. Initially, most of these plans required multi-ministry sign-off, in an attempt to require a multi-ministry (and theoretically integrated) approach. However, government reorganization and a failure to commit to a long-term adequately resourced planning process led to a government decision that each should be signed off by a single ministry. Planning and management challenges have included: difficult access to the area; the fact that the required plans have no legislatively defined content requirements or legal relationship to each other; land claims that largely remain unresolved and will take precedence over any previous planning; strong pressure from the oil and gas industry to initiate exploration and development based on high commodity prices – often before the planning has been completed; and trapping, guide-outfitting and tourism. The last of these all require high wilderness and wildlife values, but these are inadequately protected, given planning and management requirements with little understanding of cumulative effects and acceptable limits of change for any sector.

Positive aspects have included the hard work of government staff and a public board with the ability to comment on developments and make recommendations to the government about regulations, legislation and actions, and the use of a trust fund to complement government base funding for the area. Some planning has been completed, but much of this has been hampered by: heavy time pressure (both politically and from the resource sector) to act faster than is possible and thus come up with inadequate and short-term solutions; decreasing staff and financial resources; the failure of the legislation to clearly define the planning requirements and relationships between plans; and missing or delayed planning processes that would have been powerful tools to inform and assist the required plans. Each of these factors reflects the need for a well-legislated, long-term and adequately funded approach with provision for as much continuity and certainty as possible to expedite the integrated effort.

Mitchell-Banks' chapter describes how a highly ambitious project has fallen short of the mark because of many deficiencies in effective integration. Legislation is not effective in that requirements are not clearly laid out and, as a result, policy and planning are confused and conflict-laden, and there is frustration about delay and failure to accomplish the original goals as effectively as would have been possible with a coordinated and integrated approach. Nevertheless, there have been a number of successes. This case study demonstrates the need to strategically plan ahead and tactically implement integrated research and management projects, which should constantly evolve through adaptive planning and management.

White and Fisher's chapter on the development and implementation of the 1997 Banff National Park Management Plan continues the focus on the Canadian Rocky Mountains, and addresses landscape-level management in Banff National Park – a setting of evolving ecosystem and national park management paradigms. Management of the park has gone through three phases:

1   Tourism development and resource husbandry (1880s–1960), in which tourism initiatives were promoted, and fire suppression, timber harvesting (to support fire management), predator control, and culling of animal populations such as elk were the major management tools.
2   Natural regulation (1960s–1980s), with a 'bottom-up' or more 'natural' system, in which fire suppression was reduced, elk culling stopped and a general appreciation for ecosystem function and natural processes was embraced.
3   Ecological integrity and the long-term range of variability (after 1990), in which the main management aim is to maintain or restore ecological

integrity, based on three main principles: first, current ecosystems are the product of past conditions and processes; second, spatial and temporal variability in disturbance regimes are a vital attribute of ecosystems; and third, maintenance or restoration of long-term ecosystem states and processes will conserve biodiversity.

Ecological integrity and ecosystem management encourage broader citizen participation at a time when academic research is arguably becoming increasingly specialized and interest groups are becoming increasingly polarized. Reviews of the Banff National Park experience suggest a four-stage process:

1   It is critical to develop collaborative processes with stakeholders with a broad range of interests.
2   The groups of stakeholders need to be supported with a wide range of knowledge.
3   The groups need to provide decision-makers with recommendations developed through a collaborative, social learning process;.
4   Managers should adaptively implement actions with feedback from both researchers and stakeholders.

White and Fisher argue that integrated research and management require a number of steps, all requiring effective communication in a situation that is complex and continually evolving and in which many of the stakeholders may have deeply held emotional, cultural or economic stakes. The roles of common sense and mutual respect cannot be overstated.

Johnston and Good's chapter on the integrated restoration and rehabilitation of powerline corridors in Australia's national parks describes a project in which a steering committee was established with membership from various disciplines and project partners. The project incorporated specific initiatives to foster integrated approaches, including:

* the preparation of an initial scoping plan and detailed site rehabilitation plans;
* the appointment of a project manager;
* interdisciplinary stakeholder field inspections and planning days;
* structured and planned collaboration between the various parties in implementing the work plan;
* a planned and engendered ownership of the project by all stakeholders and people involved;
* planned and structured collaborative input to and preparation of all plans, training programmes, workshops, reports, manuals, and technical and scientific publications;
* establishment of an electronic 'rehabilitation network' to encourage all interested parties to discuss and contribute to the project;

- interdisciplinary workshops;
- extensive use of GIS providing a flexible capacity to integrate, manipulate and map information and data from all stakeholders, specialists and participants;
- establishment of a long-term monitoring system to assess the success of the rehabilitation projects.

The chapter describes a highly successful approach to addressing a multi-party and multi-site challenge. Vitally, 'communication' was achieved by first establishing a multidisciplinary steering committee with representation from all the relevant parties, and then staffing the team with people from different disciplines from various stakeholders, with a real focus on team building and support through meetings, plan development, information and data manipulation, the sharing of strategies, and the establishment of a long-term monitoring system to assess success. In many ways, this goes beyond the case study presented by Quinn et al, also working across jurisdictional boundaries.

Fagre, Peterson and McKenzie's chapter on the CLIMET project in the mountains of the western US, recognizes that integrated research is needed to address most issues in the science of climate change. This was accomplished through the following steps:

- developing a cadre of scientists, students and others with a common interest in the effect of climate change on natural resources;
- collecting and synthesizing existing data on climate, biology and hydrology from the three CLIMET regions;
- employing appropriate modelling techniques to synthesize data and quantify the effects of climate change on natural resources;
- applying the research results and inferences to specific resource issues by working with land managers and disseminating information to stakeholders.

As with the Swedish Mistra programme, one of the challenges of the interdisciplinary work was to maintain effective communication among all those involved. To do this, a successful strategy of workshops, conference calls and email was used. All of these placed a continued emphasis on the primary project objectives, following consistent study protocols and meeting deadlines.

Synthesis of the chapters

The 13 principles reoccur throughout the chapters of the book. At times, the principles are directly incorporated into the design and implementation of the research and/or management activities. At other times, the principles are identified as not having effectively been incorporated or missing all together, either in previous research/management work done or in the actual project or

Table 15.1 *Occurrence of the 13 principles in each chapter*

Author(s) of chapter	Decide on objectives together	Build up mutual trust	Share info, develop networks	Share responsibility	Create transparency	Monitor/evaluate collaboration	Disseminate the results	Apply the results	Share profits (results) equitably	Increase research capacity	Build on the achievements	Incorporation of TEK/local knowledge	Integrate into legislation and policy
Messerli and Messerli	X	x	X	x	X	X	X	X	x	X	X	X	
Ramakrishnan	X	X	X	X	X	X	X	X	X	X	X	X	X
Mattson and Merrill	X	x	X	X	X	X	X	x	x	X	X	X	
Brown and Schreier	X	x	X	X	X	x	X	x	X	X	X		
Mowo et al	X	X	X	X	X	X	X	X	X	X	X	X	X
Kaihura	X	X	X	X	X	X	X	X	X	X	X	X	X
Willebrand	X	X	X	X	x	X	x	X	X	x	X	X	X
Mitchley et al	X	X	X	X	X	X	X	X	X	X	x	X	
Quinn et al	X	X	X	X	X	X	X	x	x	x	X		
Mitchell-Banks	X	X	X	X	X	X	X	X	X	x	X	x	X
White and Fisher	X	X	X	X	X	X	X	x	X	X	X		
Johnston and Good	X	X	X	X	X	X	X	X	X	X	X	X	
Fagre et al	X	X	X	X	X	X	X	X	X	X	X	X	x

Note: A small 'x' indicates that the principle played a minor role or was simply referred to, while a large 'X' indicates that the principle played a major role or was explicitly referred to. An empty cell indicates no reference to a principle.

programme being discussed. Table 15.1 indicates where the principles occur in each chapter.

Complex, problem-oriented issues have to be addressed in an integrated fashion, with a strong focus on communication and coordination. Interaction between researchers and stakeholders should be as frequent and open as possible, and from the earliest possible stage – ideally when the project is being designed or the proposal is being written. There needs to be a conscious 'space' or continued opportunity for the various participants to share, discuss, argue, agree and move forward together – and this will facilitate integrated project outcomes.

Attracting funding for integrated research and producing high-quality outputs often presents many challenges, not only with regard to traditional academic benchmarks such as peer-reviewed publications, but also in terms of outputs that are policy-relevant and useful for stakeholders. The development and implementation of integrated research requires a strategy that, first, recognizes that much of academia and scientific research is managed in a single or silo discipline paradigm and, second, supports the careers of the researchers, scientists and academics while also promoting integrated research. Contributions to the research project in terms of funds and or time/expertise need to be openly documented, discussed and recognized in a transparent fashion – this, interestingly, was the synthesizing principle that was least referred to in the chapters.

Good integrated research involves effective participation and the cross-fertilization and integration of thinking, evaluation and decision-making and actions. Such research also needs to be based on a recognition of the need to understand and appreciate the various disciplinary perspectives and to essentially have a 'common language' or understanding. Good communication will ensure that a project can stay focused on achieving a successful outcome, while still being flexible to accommodate unexpected findings and understandings.

To conclude, effective integrated research and management combines: common sense; developing trust and respect through frequent communication and coordination tools or events; developing an appreciation of the role of single disciplines in creating an interdisciplinary milieu; following good practices that reflect previous lessons and proven approaches that also are tailored or amended to reflect the specific challenges at hand; ensuring the timely participation of both stakeholders and researchers in cooperatively working together to address complex and often 'wicked' problems; and, once these problems have been addressed, sharing and disseminating the results and lessons not only among researchers and stakeholders but also with the larger academic, government, industry and general public audiences. Finally, this shared information and experience should also develop research and management capacity, be applied 'on the ground', and contribute to effective policy and legislation.

References

KFPE (Swiss Commission for Research Partnership with Developing Countries) (2003) *Guidelines for Research Partnership with Developing Countries*, 3rd edn, Berne, KFPE

Index